# Engineering Applications of FPGAs

Esteban Tlelo-Cuautle
José de Jesús Rangel-Magdaleno
Luis Gerardo de la Fraga

# Engineering Applications of FPGAs

## Chaotic Systems, Artificial Neural Networks, Random Number Generators, and Secure Communication Systems

Esteban Tlelo-Cuautle
INAOE
Tonantzintla, Puebla
Mexico

José de Jesús Rangel-Magdaleno
Department of Electronics
INAOE
Tonantzintla, Puebla
Mexico

Luis Gerardo de la Fraga
Computer Science Department
CINVESTAV-IPN
Mexico City
Mexico

ISBN 978-3-319-81679-1 ISBN 978-3-319-34115-6 (eBook)
DOI 10.1007/978-3-319-34115-6



Printed on acid-free paper

This Springer imprint is published by Springer Nature
The registered company is Springer International Publishing AG Switzerland

*The authors want to dedicate this book to their families for the support provided during the preparation of this material.*

# Preface

Field-programmable gate arrays (FPGAs) were invented in 1984 by Ross Freeman. Basically, it is a semiconductor consisting of programmable logic blocks that can be used to reproduce simple functions up to a complex system on a chip (SoC). The main advantages of the FPGAs are: they can be reprogrammed, have low development and acquisition costs, and their application is a good option if the product is not in high numbers. That way, FPGAs are gaining the attention of researches for the development of applications in a wide variety of fields, for example, medicine, communications, signal processing, avionics, space, finance, military, electronics, and other areas that exploit their flexibility and capability of being reprogrammed/configured.

Configurability for engineering applications makes FPGA very crucial in initial stages for any embedded project. Some analog circuits and any digital circuit can be implemented using FPGA, so the possibilities are endless. However, applications found on recent articles and books did not detail the realizations from the model to the physical synthesis. That way, this book details engineering applications of FPGAs from mathematical models descriptions to VHDL programming issues and hardware implementation of applications involving chaos theory.

The reader can find insights on FPGA-based implementations for chaos generators, artificial neural networks (ANNs), random number generators (RNGs), and master–slave synchronization of chaotic oscillators to implement a secure communication system for image transmission. The plus of this book is focused on providing VHDL programming guidelines and issues, along with co-simulation examples with Active-HDL and Simulink. In addition, we list some challenges on applying different kinds of numerical methods, problems on optimizing chaotic systems, selection of an ANN topology, its training, improvements on designing activation functions, data supply using serial communication with a computer, generation of random number generators from chaos theory, realization of chaotic secure communication systems, and other open problems for future research.

In summary, this book details FPGA realizations for:

- Chaos generators, which are described from their mathematical models, are characterized by their maximum Lyapunov exponent, and are implemented using minimal FPGA resources.
- Artificial neural networks (ANNs), discussing some topologies, different learning techniques, kinds of activation functions, and issues on choosing the length of the digital words being processed. One ANN topology is applied to chaotic time series prediction.
- Random number generators (RNGs), which are designed using different chaos generators, in the continuous-time and discrete-time domains. The RNGs are characterized by their maximum Lyapunov exponent and entropy, and are evaluated through NIST tests.
- Optimized chaotic oscillators are synchronized in a master–slave topology that is used to implement a secure communication system to process black and white, and grayscale images.

Some chapters discuss computer arithmetic issues to minimize hardware resources and to reduce errors, before synthesizing the FPGA realization. At the end, the reader can infer open lines for future research not only in areas where chaos generators, ANNs, random number generators, and secure communications are required, but also to extend the presented material to other problems in engineering.

# Acknowledgments

The authors acknowledge the pretty good help of the graduate students at INAOE for preparing simulations and experiments. They are: Ana Dalia Pano-Azucena and Antonio de Jesus Quintas-Valles.

Special acknowledgments to CONACyT-Mexico for funding support under projects 168357 and 237991.

# Contents

# Acronyms

| | |
|---|---|
| ALM | Adaptive logic module |
| ANN | Artificial neural network |
| ASIC | Application-specific integrated circuit |
| CAD | Computer-aided design |
| CLB | Configurable logic block |
| CLK | Clock |
| CLT | Central limit theorem |
| CRAM | Configuration of random access memory |
| CTW | Context-tree weighting |
| DCM | Digital clock manager |
| DSP | Digital signal processor |
| EDA | Electronic design automation |
| FPGA | Field-programmable gate array |
| FPL | Field-programmable logic |
| FSM | Finite-state machine |
| GNG | Gaussian noise generator |
| HDL | Hardware description language |
| IP | Intellectual property |
| LC | Logic cells |
| LE | Logic elements |
| LR | Learning rule |
| LSB | Least significant bit |
| LUT | Look-up table |
| MC | Moment constant |
| MEMS | Microelectromechanical system |
| MLE | Maximum Lyapunov exponent |
| MSB | Most significant bit |
| MSE | Mean square error |
| NIST | National Institute for Standards and Technology |
| ODE | Ordinary differential equation |
| OPC | Operation counter |

| | |
|---|---|
| PC | Personal computer |
| PLL | Phase-locked loop |
| PWL | Piecewise linear |
| RAM | Random access memory |
| RNG | Random number generator |
| ROM | Read-only memory |
| RST | Reset |
| RTL | Register transfer level |
| SCM | Single constant multiplication |
| SFS | Saturated function series |
| SM | Sign-magnitude |
| SQNR | Signal to quantization noise ratio |
| SRAM | Static random access memory |
| TDL | Tapped delay line |
| VHDL | Very-high-speed integrated circuit hardware description language |

# Chapter 1
# Introduction to Field-Programmable Gate Arrays

## 1.1 FPGA Architectures

What is an FPGA? Field-programmable gate arrays (FPGAs) are a class of devices classified or called as field-programmable logic (FPL).

FPGAs are programmable semiconductor devices that are based on a matrix of configurable logic blocks (CLBs) connected through programmable interconnects [1]. Contrary to application-specific integrated circuits (ASICs), which have a fixed design based on the application and that cannot be modified, FPGAs can be reconfigured to modify or improve the design according to the application requirements.

The FPGA is a device that is completely manufactured, but that remains design-independent allowing flexibility in implementing designs. Each FPGA vendor designs a reconfigurable architecture, for example based on CLBs, logic cells (LCs) or logic elements (LEs), provided by Xilinx and Altera, respectively. Also the FPGAs include hard blocks like random access memory (RAM) or digital signal processor (DSP), which are commonly used devices, while their incorporation improves the resource utilization and maximum frequency of operation.

Figure 1.1 shows a simplified structure of a FPGA provided by Xilinx. One can see that in general three major types of elements are required [2] as follows:

- Logic blocks
- I/O blocks
- Programmable Interconnect

Today, depending on FPGA's family, its architecture may include many additional hardware components that are integrated directly into the FPGA fabric, such as embedded multipliers or DSP blocks, RAM blocks, digital clock managers (DCMs), phase-locked loops (PLLs), soft processors, and intellectual property (IP) cores.



E. Tlelo-Cuautle et al., *Engineering Applications of FPGAs*,
DOI 10.1007/978-3-319-34115-6_1

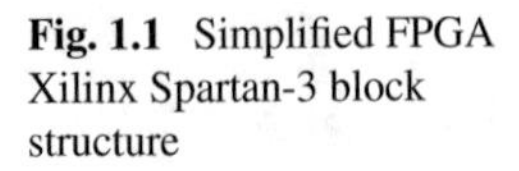

**Fig. 1.1** Simplified FPGA Xilinx Spartan-3 block structure

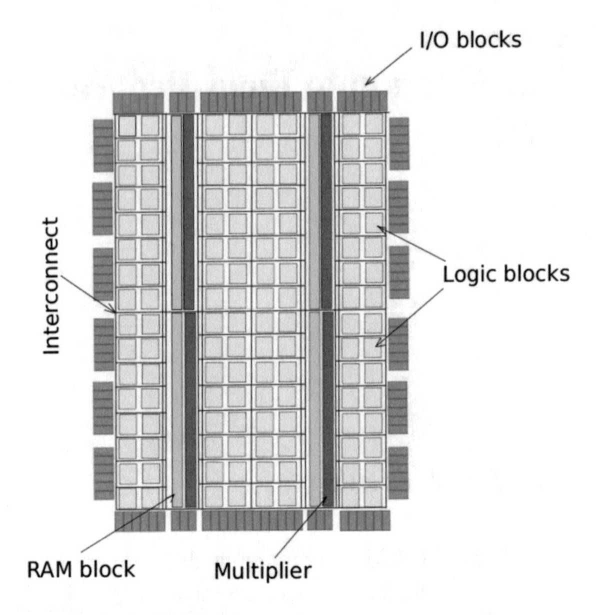


## 1.2 Blocks Description

### *1.2.1 Logic Blocks*

Static memory is the most widely used method of configuring FPGAs. Logic blocks are based on this method; Xilinx called their logic blocks as **configurable logic block (CLB)** and Altera called **adaptive logic module (ALM)**. To better understand how the logic blocks are constituted, a general description of CLBs and ALMs is presented below.

#### 1.2.1.1 Xilinx CLB

Talking about Xilinx, CLBs constitute the main logic resource for implementing synchronous and combinational circuits. Taking into account Xilinx Spartan-3 Family, each CLB contains four slices (these slices are grouped in pairs, each pair is organized as a column with an independent carry chain), an interconnect routing to neighboring CLBs and a switch matrix to provide access to general routing resources as shown by Fig. 1.2 [3].

All the slides into a CLB contain the following elements: two RAM-based function generators (also known as a lookup table (LUT), is the main resource for implementing logic functions), two storage elements, wide-function multiplexers, two independent carry logic (which runs vertically up only to support fast and efficient implementations of math operations), and arithmetic gates as shown in Fig. 1.3 [3].

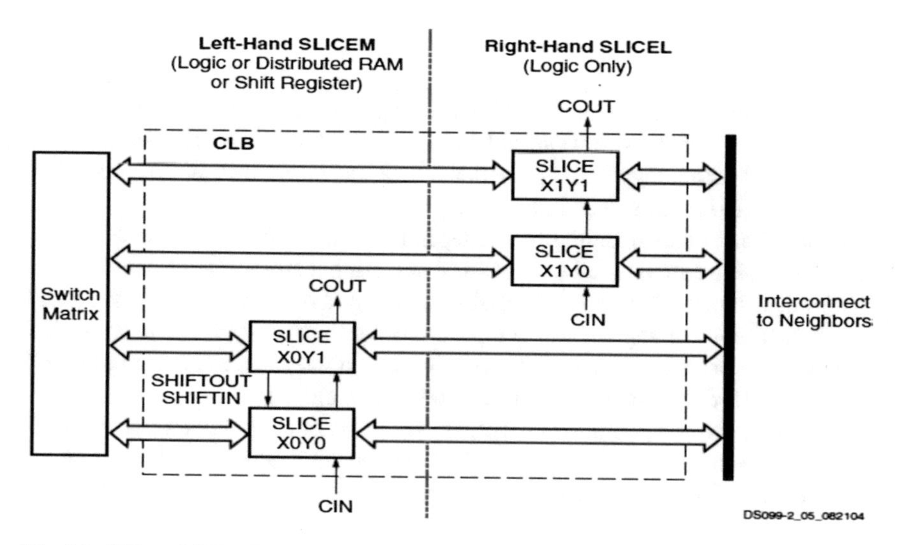


**Fig. 1.2** Slides of CLB (© 2013 Xilinx)

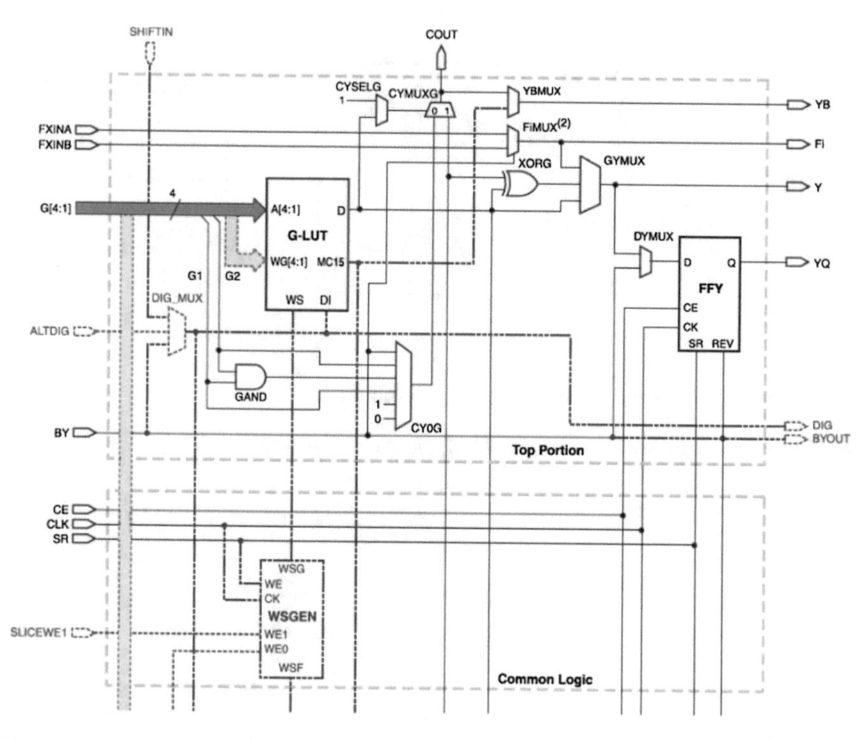


**Fig. 1.3** Spartan-3 low portion of a slide (© 2013 Xilinx)

#### 1.2.1.2 Altera ALM

Altera's adaptive logic module (ALM) technology consists of eight-input combinational logic, two registers, and two adders as shown in Fig. 1.4 [4]. By using Altera's patented LUT technology the combinational logic can be divided between two adaptive LUTs (ALUTs) as shown in Fig. 1.5.

In addition to implementing two independent 4-input functions, the ALM can for example, implement a full 6-input LUT or a 5-input and a 3-input function with independent inputs. Because two registers and two adders are available, the ALM has the flexibility to implement 2.5 logic elements (LEs) of a classic 4-input LUT architecture, consisting of a 4-LUT, carry logic, and a register as shown in Fig. 1.6 for Cyclone III device family in normal mode operation [5].

Normal mode operation allows general logic applications and combinational functions implementation. The Cyclone III LES can be configured in arithmetic mode, which is ideal for implementing adders, counters, accumulators, and comparators.

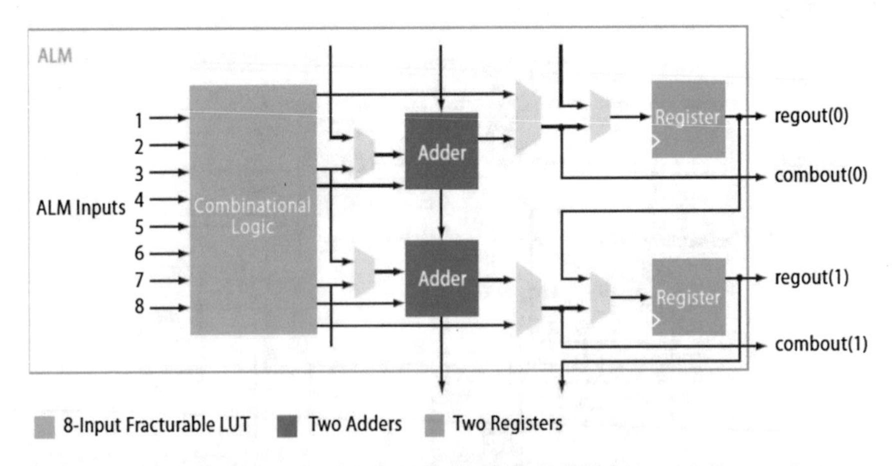


**Fig. 1.4** Altera's adaptive logic module (ALM) block diagram (© 2006 Altera)

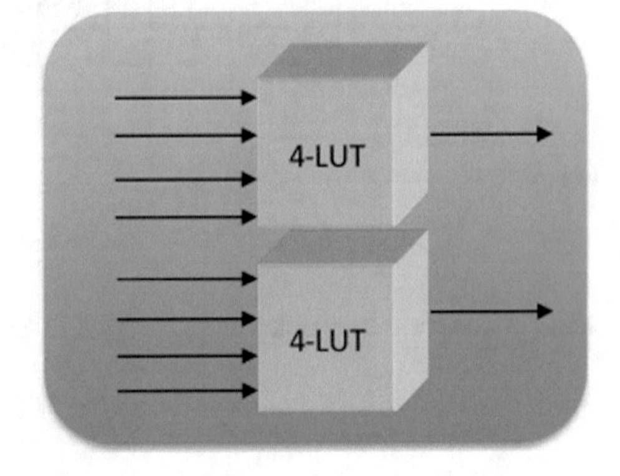


**Fig. 1.5** Two independent 4-input LUTs (© 2006 Altera)

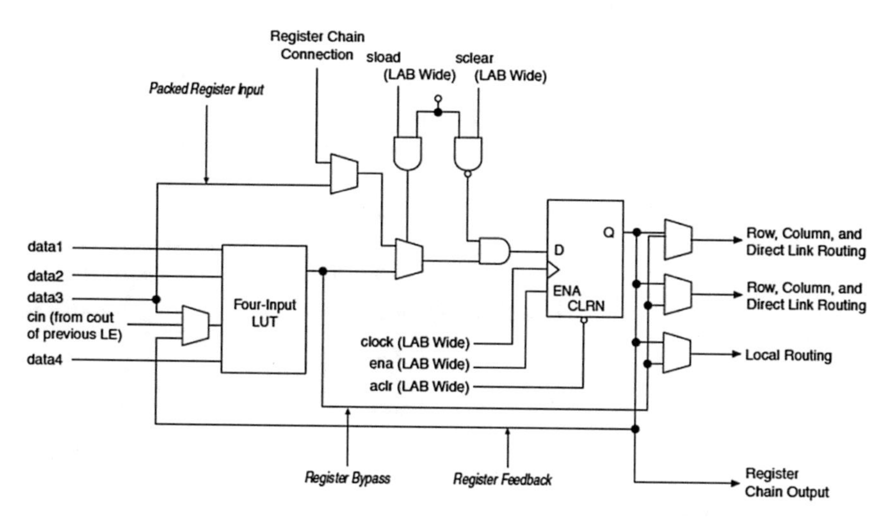


**Fig. 1.6** Cyclone III device family LEs (© 2011 Altera)

The Quartus II Compiler automatically selects the operation mode during design processing. For normal mode operation the Quartus II Compiler selects the carry-in signal as one of the inputs to the LUT, while for arithmetic mode creates carry chain logic runs vertically, which allows fast horizontal connections.

## *1.2.2 Lookup Tables (LUTs)*

Look-Up Tables are a fundamental part in logic elements of both Xilinx's CLB and Altera's ALM providers.

### 1.2.2.1 Xilinx LUTs

The LUT is a RAM-based function generator and is the main resource for implementing logic functions. Furthermore, the LUTs in each slice pair can be configured as distributed RAM or a 16-bit shift register.

Each of the two LUTs (F and G) in a slice has four logic inputs (A1–A4) and a single output (D) (see Fig. 1.7). Any four-variable Boolean logic operator can be implemented in one LUT. Functions with more inputs can be implemented by cascading LUTs or by using the wide-function multiplexers. The output of the LUT can connect to the wide multiplexer logic, the carry and arithmetic logic, or directly to a CLB output or to the CLB storage element [6].

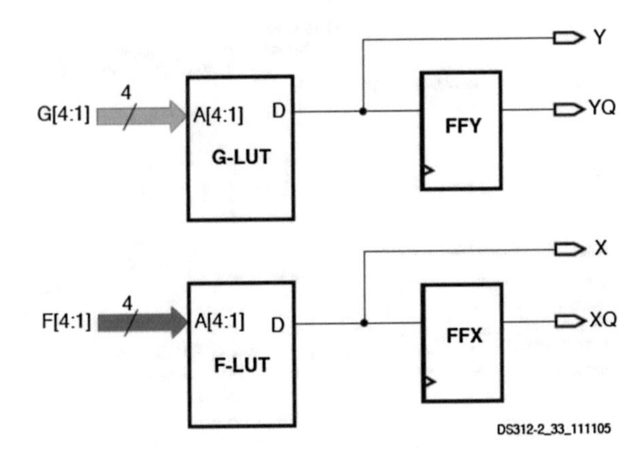


**Fig. 1.7** Xilinx LUTs (© 2011 Xilinx)

#### 1.2.2.2 Altera LUTs

An Altera LUT is typically built out of SRAM bits to hold the configuration memory (CRAM) LUT-mask and a set of multiplexers to select the bit of CRAM that is to drive the output. Figure 1.8 shows a 4-LUT, which consists of 16 bits of SRAM and a 16:1 multiplexer implemented as a tree of 2:1 multiplexers. The 4-LUT can implement any function of four inputs (A, B, C, D) by setting the appropriate value in the LUT-mask [4].

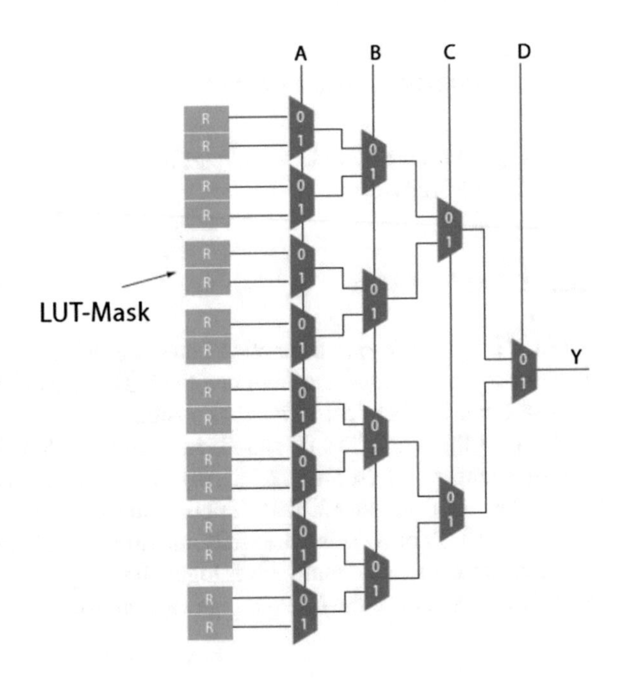


**Fig. 1.8** Altera LUTs (© 2006 Altera)

### 1.2.3 I/O Blocks

Generally I/O blocks contain registers and buffers, their interconnection depend on the FPGA family. A general description for Xilinx and Altera I/O blocks is presented below.

#### 1.2.3.1 Xilinx I/O Blocks

Xilinx I/O Blocks have three main signal paths: the output path, input path, and 3-state path. Each path has its own pair of storage elements that can act as either registers or latches (see Fig. 1.9). I/O Blocks provides a programmable, unidirectional or bidirectional interface between package pin and the FPGA's internal logic, supporting a wide variety of standard interfaces [6].

#### 1.2.3.2 Altera I/O Elements

The Altera Cyclone IV I/O Elements contain a bidirectional I/O buffer and five registers for registering input, output, output-enable signals, and complete embedded bidirectional single-data rate transfer (see Fig. 1.10). I/O pins support various single-ended and differential I/O standards [7].

## 1.3 Programming Environments

In this section a brief introduction to Vivado from Xilinx, Quartus II from Altera, and Active-HDL is presented.

### 1.3.1 Vivado

Vivado is the software tool that Xilinx provides for simulation, register transfer level (RTL) analysis, synthesis, implementation, programming and debug [8]. Figure 1.11 shows the first window of Vivado, the next options are presented as follows:

- Quick Start
  - Create New Project
  - Open Project
  - Open Example Project
- Tasks

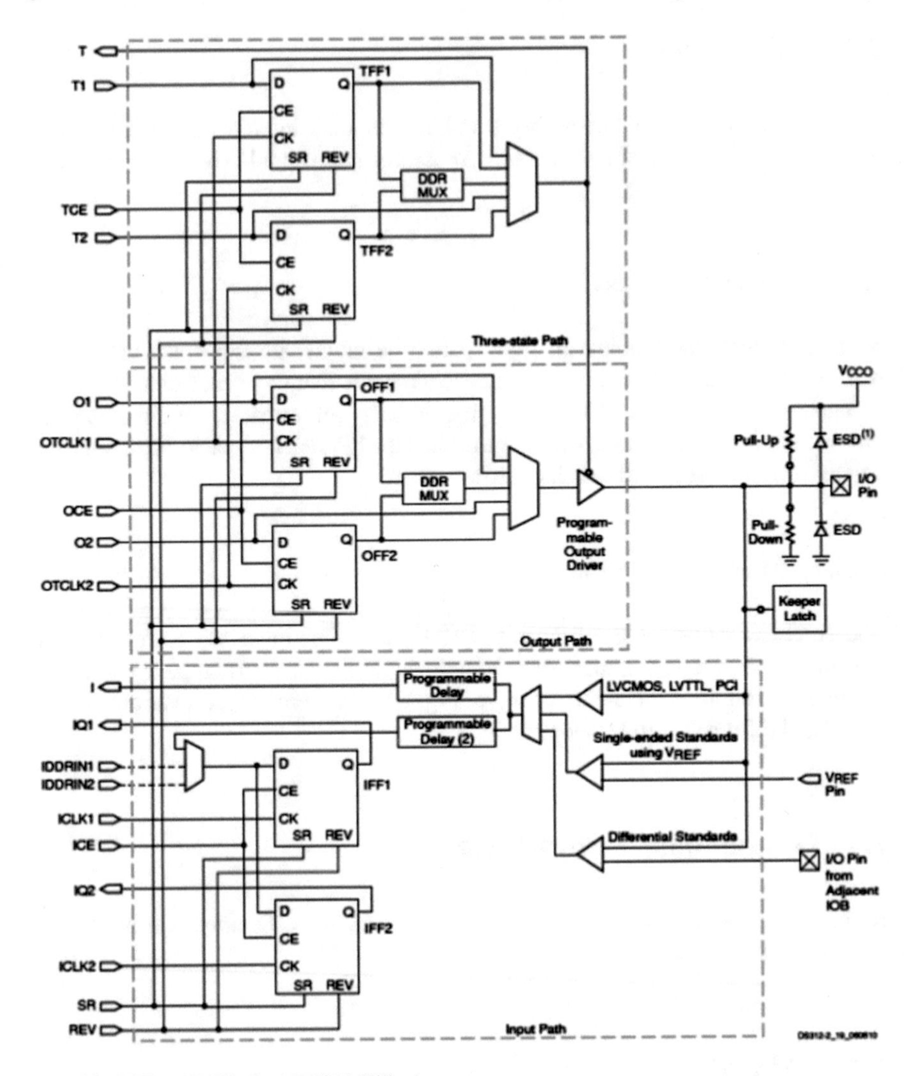


**Fig. 1.9** Xilinx I/O Blocks (©2011 Xilinx)

- Manage IP
- Open Hardware Manager
- Xilinx Td Store

- Information Center
  - Documentation and Tutorials
  - Quick Take Videos
  - Release Notes Guide

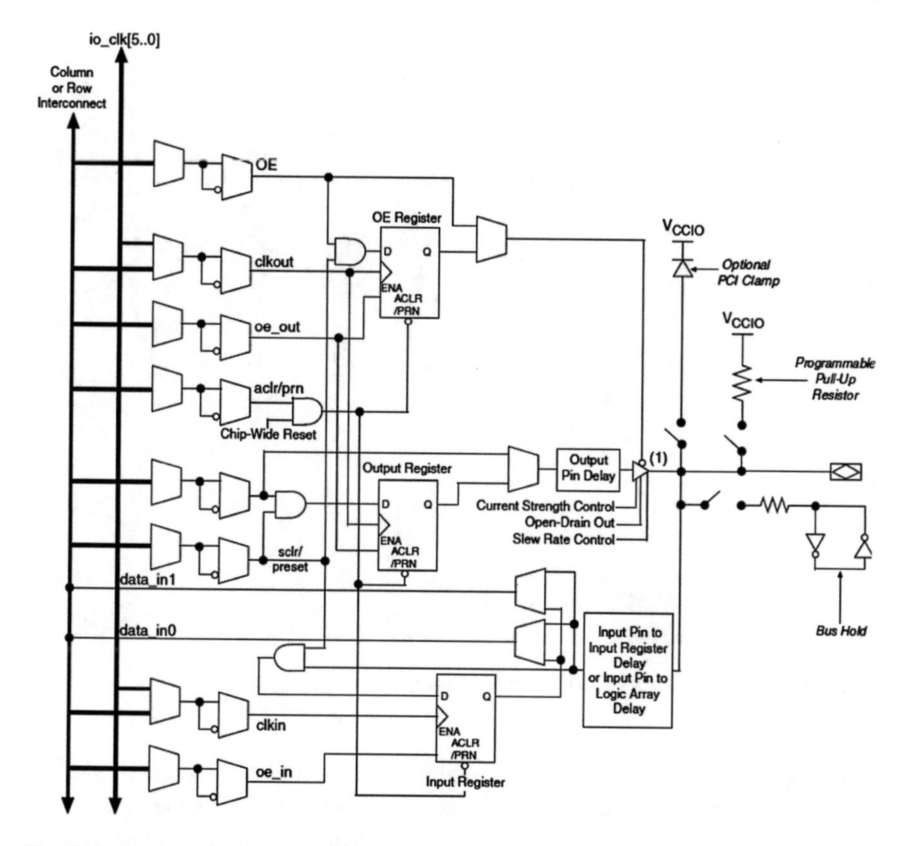


**Fig. 1.10** Altera I/O blocks (© 2013 Altera)

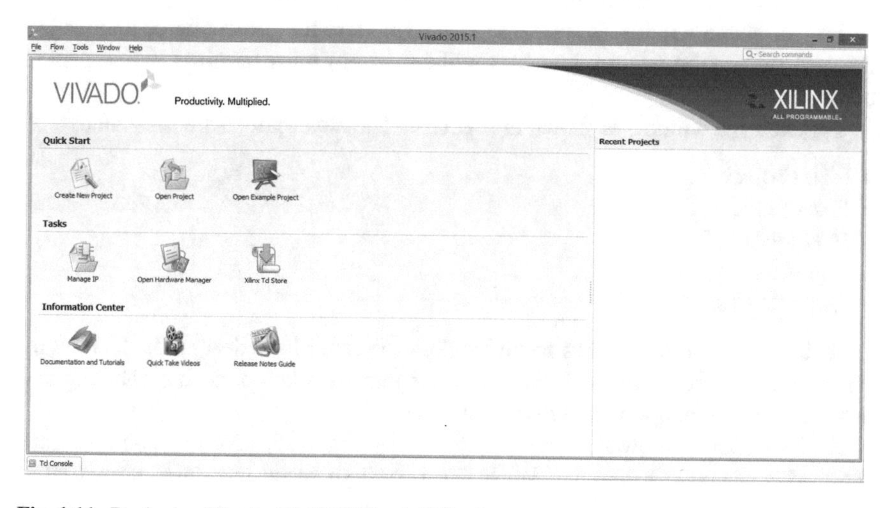


**Fig. 1.11** Beginning Vivado (© 2015 Vivado Xilinx)

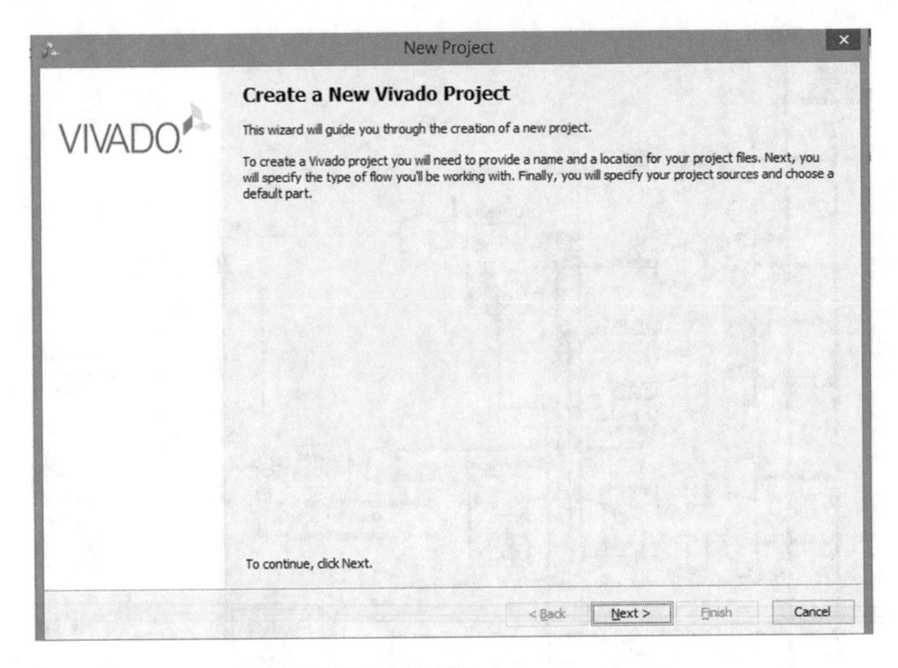


**Fig. 1.12** Create new project (© 2015 Vivado Xilinx)

If there are any recent projects they will be listed in Recent projects subwindow. By clicking create new project link, the window shown in Fig. 1.12 appears, and a wizard to create the new project is started. The instructions to create a new Vivado project are given. To continue, click next.

Project Name window is shown in Fig. 1.13. Here the project name and project location (where the project data files will be stored) must be given. Click next to continue.

Project Type window is shown in Fig. 1.14. The next options are presented:

- RTL Project
- Post-synthesis Project
- I/O Planning Project
- Imported Project
- Example Project

RTL Project option allows to add sources, create block designs in IP Integrator, generate IP, run RTL analysis, synthesis, implementation, design planning and analysis. Click this option and then click next.

Add sources window is opened as shown by Fig. 1.15, here one can add files, directories or create a new file. Verilog or very high speed integrated circuit HDL (VHDL) files can be added, also simulation language is selected. Vivado supports mixed description and simulation (Verilog and VHDL). Click next to continue.

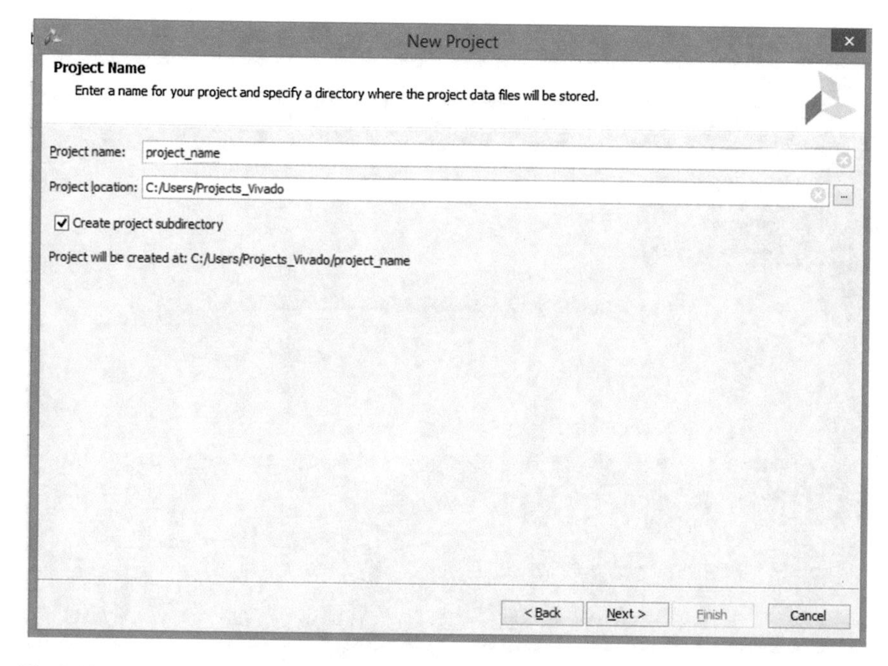


**Fig. 1.13** Project name and location (© 2015 Vivado Xilinx)

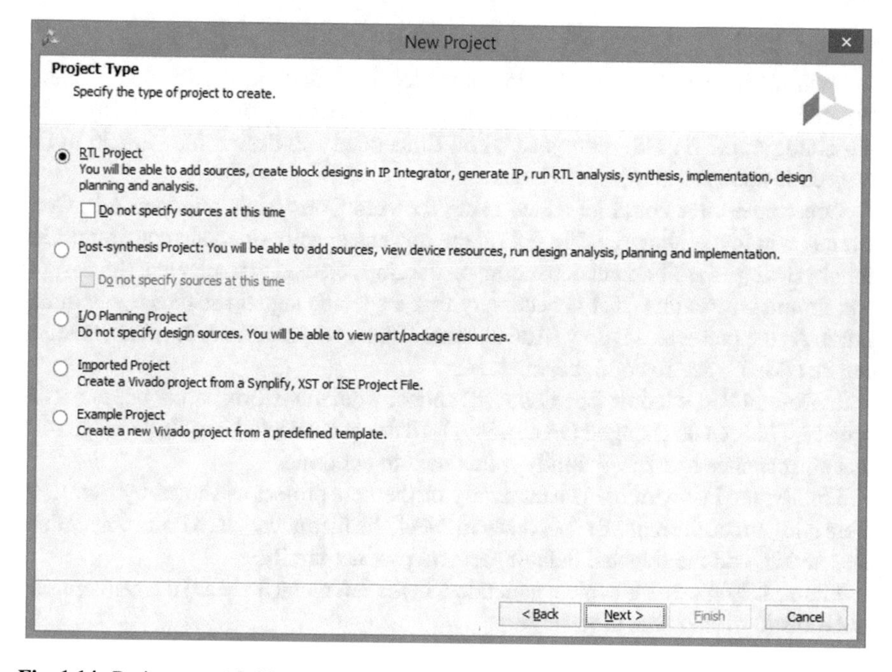


**Fig. 1.14** Project type (© 2015 Vivado Xilinx)

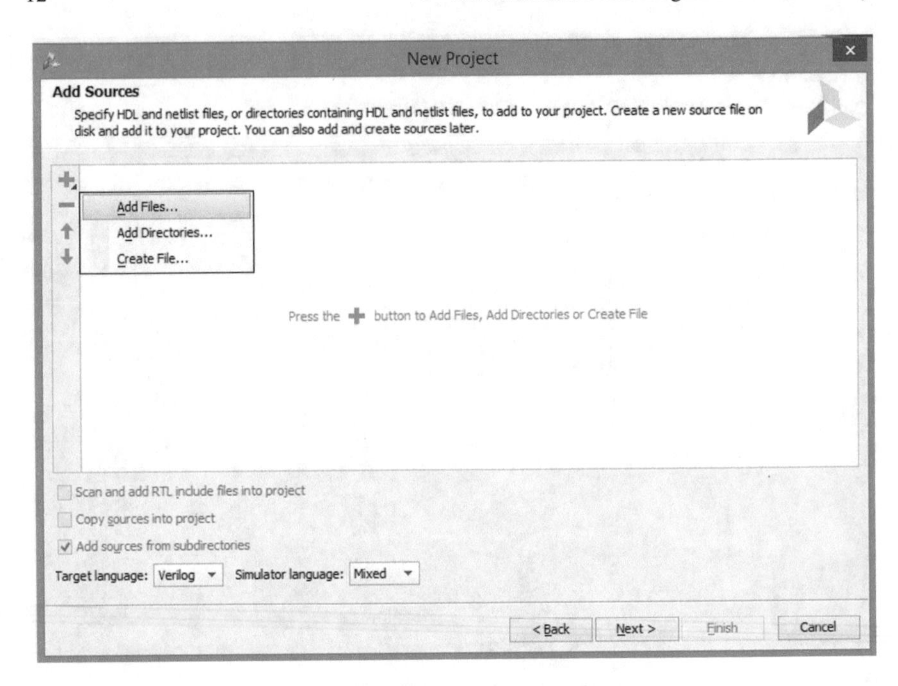


**Fig. 1.15** Add sources (© 2015 Vivado Xilinx)

With all the sources included the next step is to include the IP files, if there. Figure 1.16 shows the Add Existing IP window. This window is for specifying existing configurable IP, DSP composite, and Embedded sub-design files to add to the project. If there are not, just click next.

Constraints files could be added using the Add Constraints window. Add Constraints window is shown in Fig. 1.17, here one can specify or create constraint files for physical (pin assignment depending on the chip selected to implement the design) and timing constraints (if it is necessary that a specific signal satisfies a maximum time). As the previous window (Add existing IP), add constraints is optional, the files can be added later. To continue, click next.

In Default Part window (see Fig. 1.18) choose a default Xilinx part or board for the project. This can be changed later. In Fig. 1.18 the part selected is xc7a100tcsg324-1 that corresponds to Artix-7 family. Click next to continue.

Finally, the last window is a summary of the new project as shown by Fig. 1.19. Here one can see the name of the created project, the files added, the IP and constraints files added, and the selected default part and product family.

Figure 1.20 shows the project generated, in the left menu there are the next options and tools to manage the project:

- Project Manager
  - Project Settings

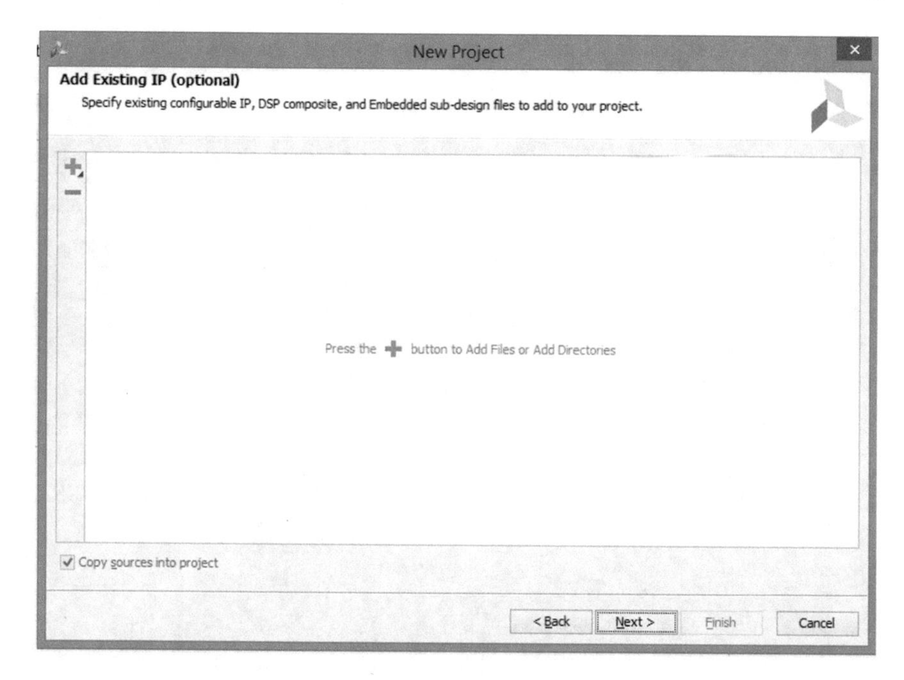


**Fig. 1.16** Add existing IP (© 2015 Vivado Xilinx)

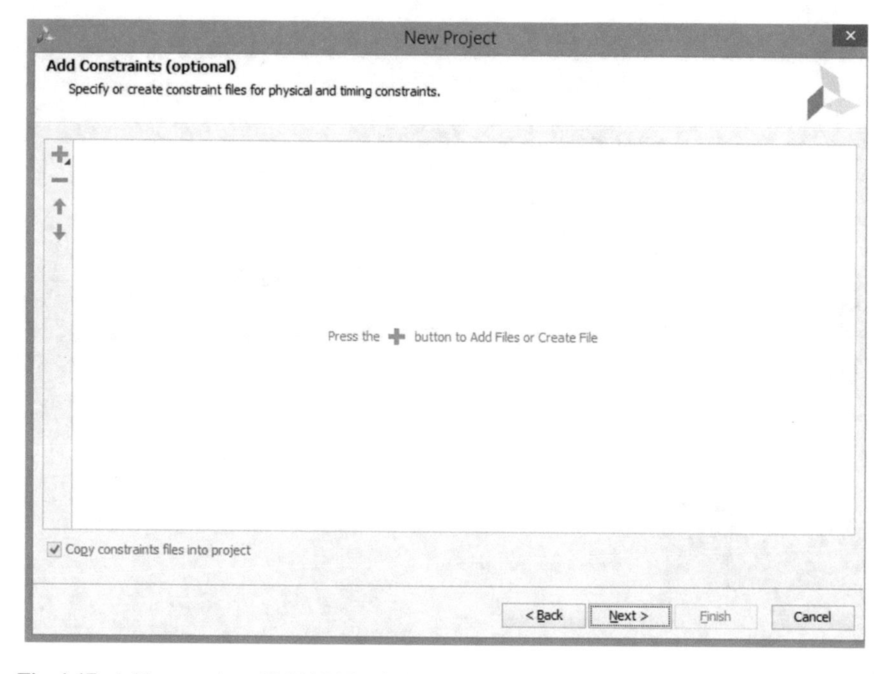


**Fig. 1.17** Add constraints (© 2015 Vivado Xilinx)

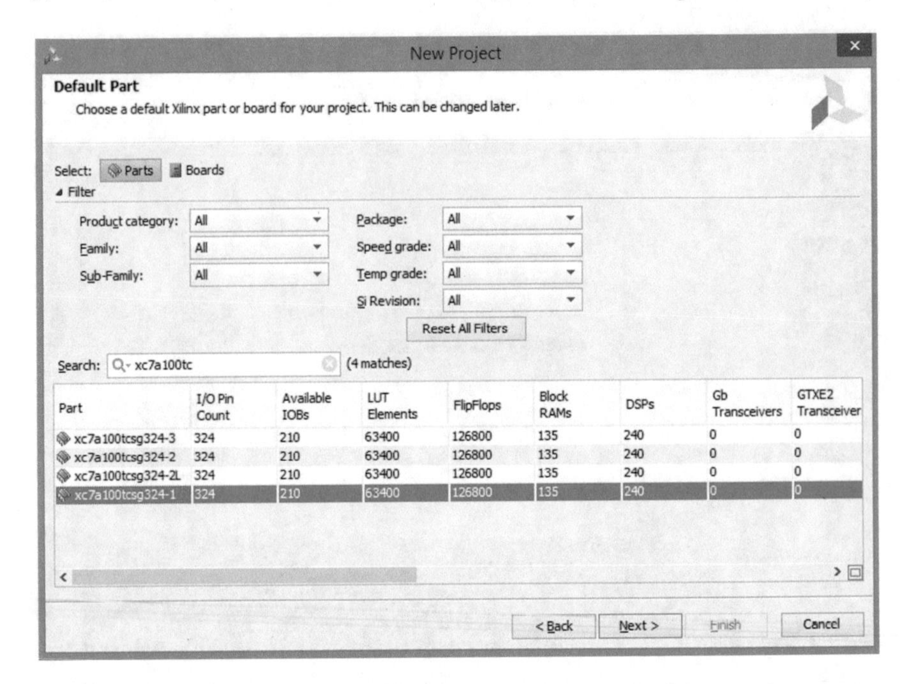


**Fig. 1.18** Default part (© 2015 Vivado Xilinx)

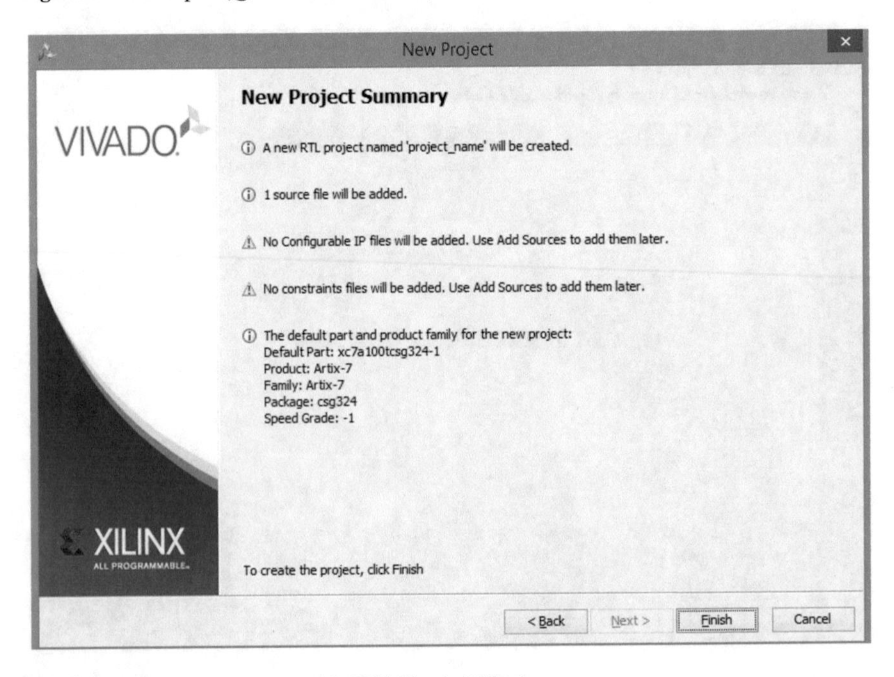


**Fig. 1.19** New project summary (© 2015 Vivado Xilinx)

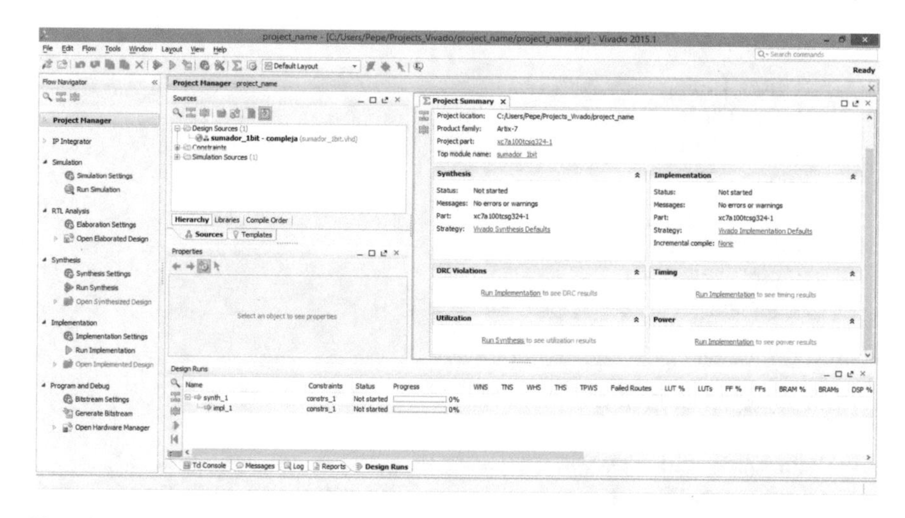


**Fig. 1.20** Project generated (© 2015 Vivado Xilinx)

  - Add Sources
  - Language Templates
  - IP Catalog

- IP Integrator
  - Create Block Design
  - Open Block Design
  - Generate Block Design

- Simulation
  - Simulation Settings
  - Run Simulation

- RTL Analysis
  - Elaboration Setting
  - Open Elaborated Design
    - Report DRC
    - Report Noise
    - Schematic

- Synthesis
  - Synthesis Setting
  - Run Synthesis
  - Open Synthesis Design
    - Constraints Wizard
    - Edit Timing Constraints

Set Up Debug
Report Timing Summary
Report Clock Networks
Report Clock Iteration
Report DRC
Report Noise
Report Utilization
Report Power
Schematic

- Implementation
  - Implementation Settings
  - Run Implementation
  - Open Implemented Design
    Constraints Wizard
    Edit Timing Constraints
    Report Timing Summary
    Report Clock Networks
    Report Clock Iteration
    Report DRC
    Report Noise
    Report Utilization
    Report Power
- Program and Debug
  - Bitstream Settings
  - Generate Bitstream
  - Open Hardware Manager
    Open Target
    Program Device
    Add Configuration Memory Device

### *1.3.2 Quartus II*

Quartus II is the software tool from Altera, Quartus II software is a complete computer-aided design (CAD) system for designing digital circuits [9]. Figure 1.21 shows the first window when Quartus II is started. In the central window (Home) there are the Start Designing and Recent Projects options. By clicking New Project Wizard the user starts the design of a new project.

Figure 1.22 shows the New Project Wizard window, in which the user selects the working directory for the project, gives a name to the project, and writes the name of the top-level entity for the project (it is important to underline that the name is case

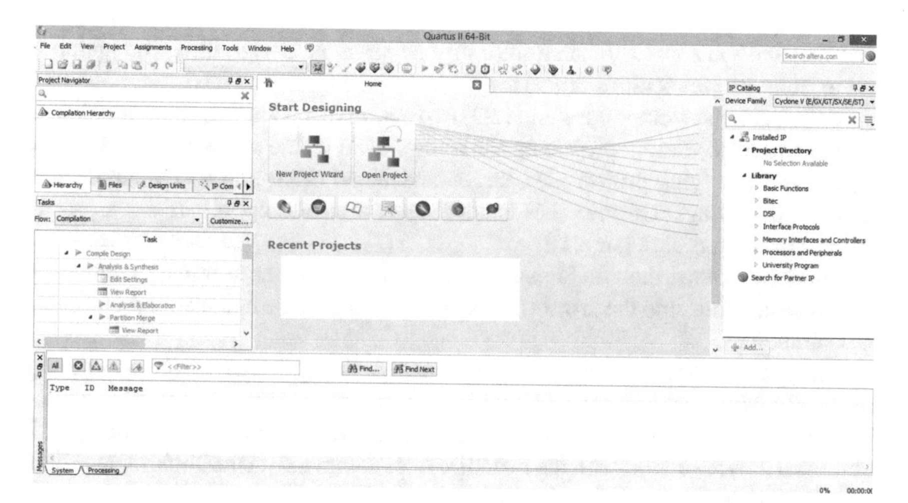


**Fig. 1.21** Beginning Quartus II (© 2015 Quartus II Altera)

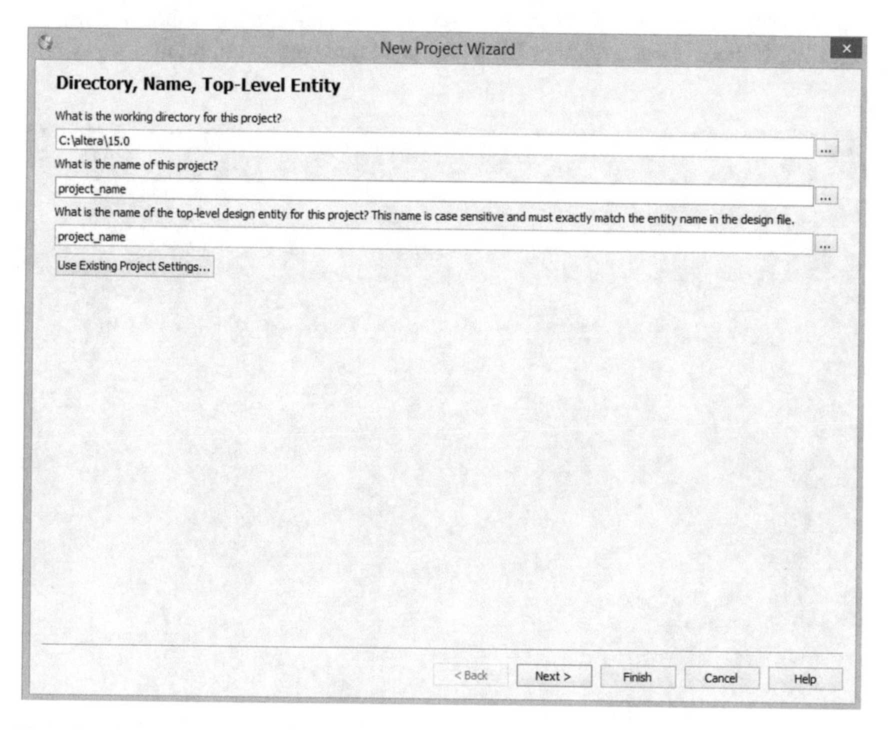


**Fig. 1.22** Quartus II new project wizard (© 2015 Quartus II Altera)

sensitive and must exactly match the entity name in the design file). The option to use existing project settings is also available. To continue click next.

In Project Type window (see Fig. 1.23) two options to create the project are shown. On the one hand, Empty project option allows us to create a new project from the beginning, specifying project files, target device and electronic design automation (EDA) tool settings. On the other hand, Project template option creates a project from an existing design template.

Figure 1.24 shows the Add Files window, in this window the user selects the design files to be included into the project. If there are not files to be included just click next to continue.

Family and Devices Settings window is shown in Fig. 1.25, here the user selects the family device and device to target for compilation. Depending on the Quartus II version there are some Family devices included, if necessary the user can install additional device support with the InstallDevices command on the Tools menu. Click next to continue.

Quartus II gives the option to specify other EDA tools to be used with Quartus II to develop the project, this can be done in EDA Tool Setting window shown in Fig. 1.26. If there is not any other EDA tool just click next to continue.

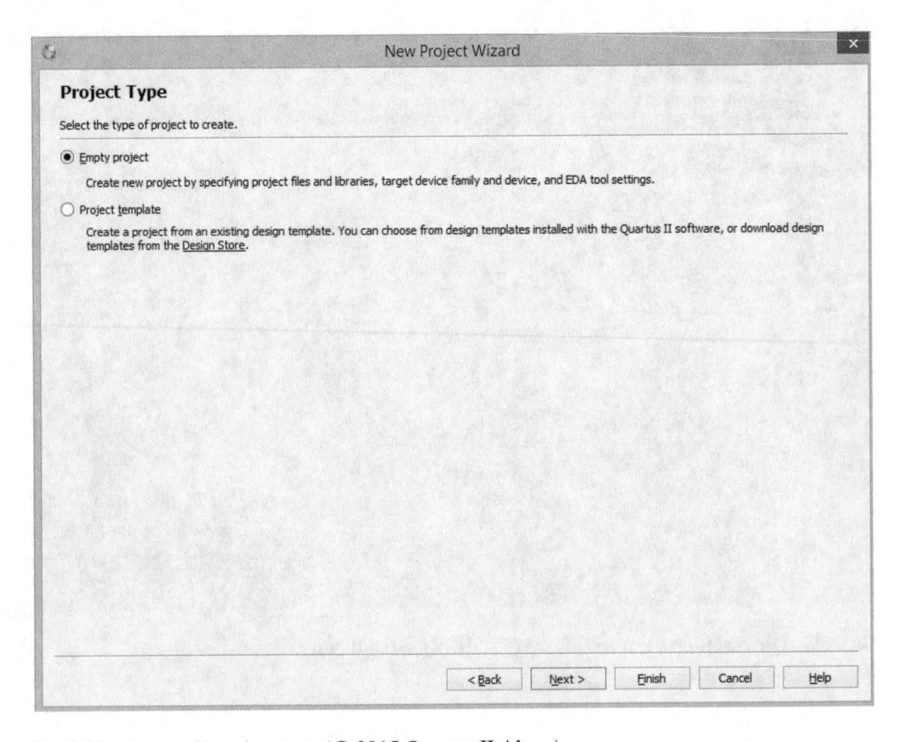


**Fig. 1.23** Quartus II project type (© 2015 Quartus II Altera)

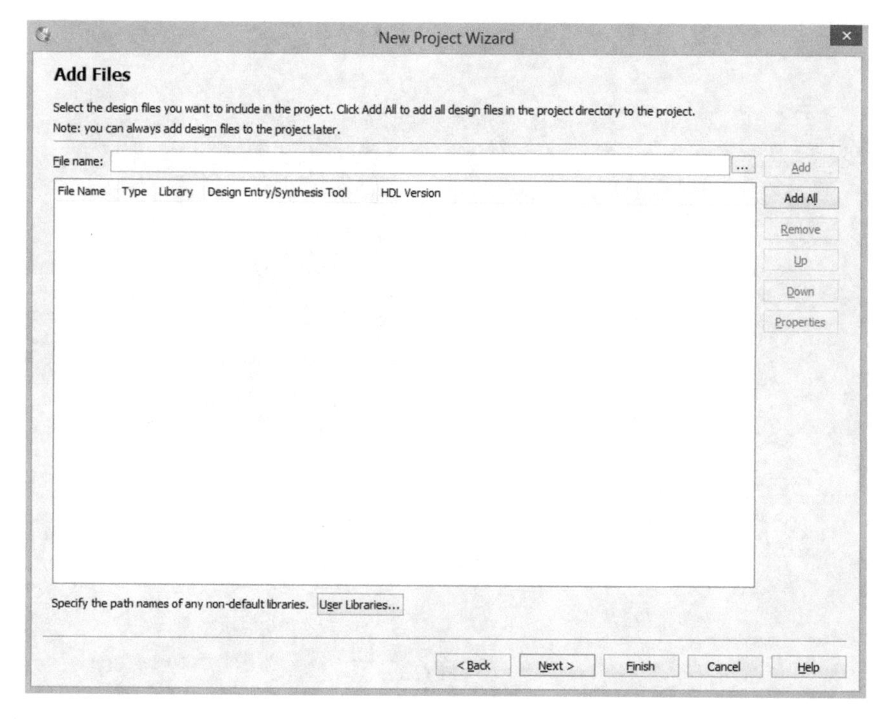


**Fig. 1.24** Quartus II add files (© 2015 Quartus II Altera)

In the last wizard window a summary of the new project created is shown (see Fig. 1.27), if necessary the user can go back to change one or several selected options, if not just click finish to generate the new project.

Finally, Fig. 1.28 shows the new project generated in Quartus II software, in the left menu the user can apply the following options for compiling the design:

- Analysis and Synthesis
  - Edit Settings
  - View Report
  - Analysis and Elaboration
  - Partition Merge
    View Report
    State Machine Viewer
    Technology Map Viewer (Post-Mapping)
  - Netlist Viewer
    RTL Viewer
    Design Partition Planner
  - Design Assistant (Post-Mapping)
    View Report
    Edit Settings

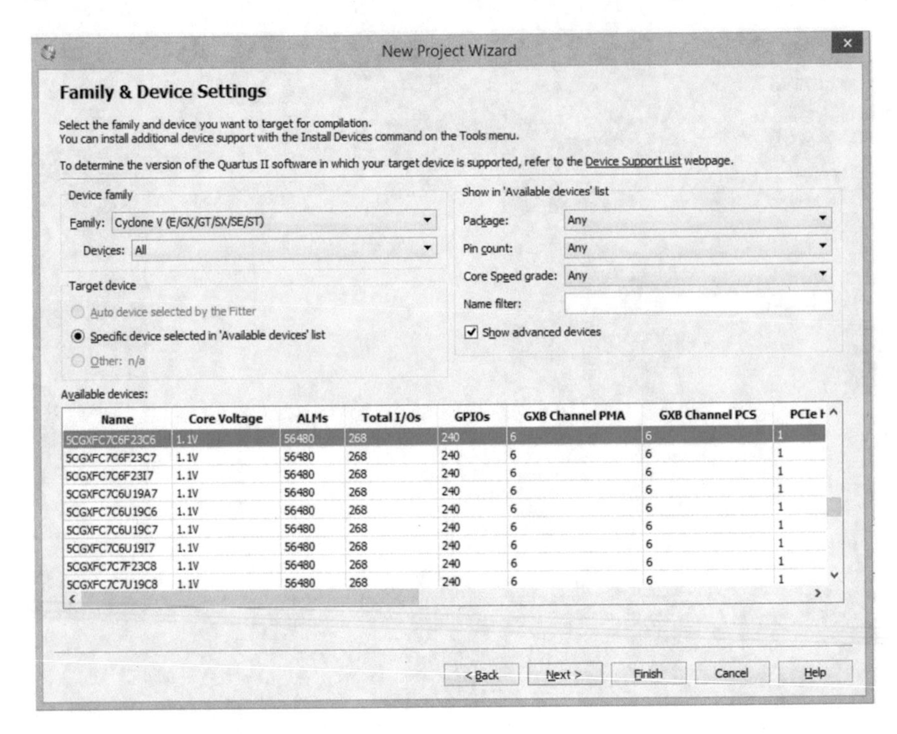


**Fig. 1.25** Quartus II family and devices settings (© 2015 Quartus II Altera)

- I/O Assignment Analysis
  View Report
  Pin Planner

- Fitter (Place and Route)
  - View Report
  - Edit Settings
  - Chip Planner
  - Technology Map Viewer (Post-Fitting)
  - Design Assistant (Post-Fitting)
    View Report
    Edit Settings

- Assembler (Generate programming files)
  - View Report
  - Edit Settings

- TimeQuest Timing Analysis
  - View Report
  - Edit Settings

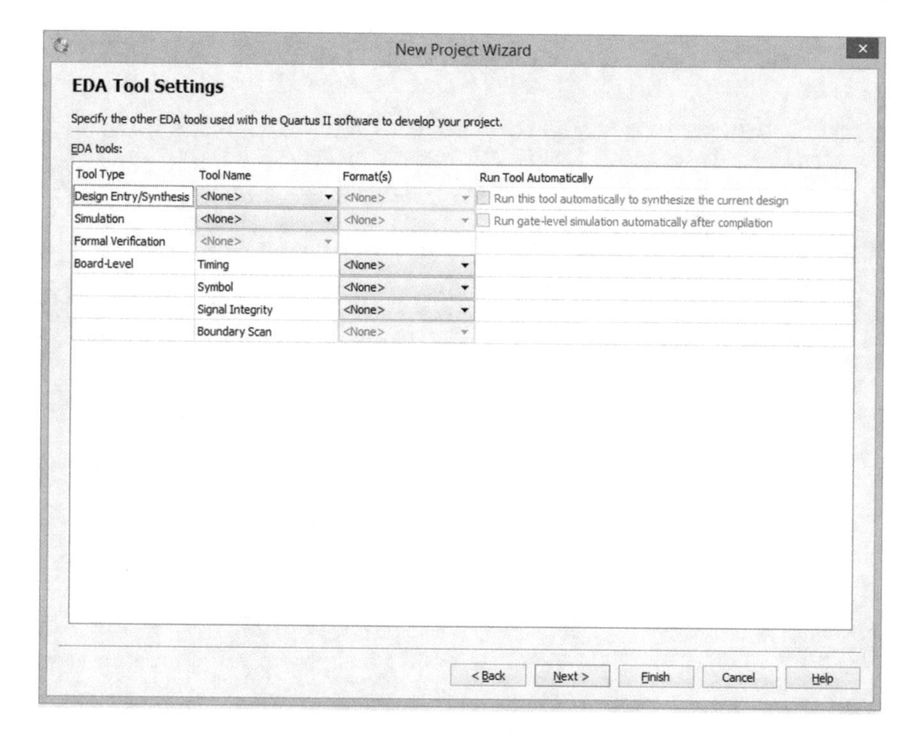


Fig. 1.26 Quartus II EDA tool settings (©2015 Quartus II Altera)

  - TimeQuest Timing Analyzer
- EDA Netlist Writer
  - View Report
  - Edit Settings
- Program Device (Open Programer)

### *1.3.3 Aldec Active-HDL*

Aldec Active-HDL is a windows based, integrated FPGA design creation and simulation solution team-based environment [10]. It is an excellent tool for digital circuit simulation described in VHDL and Verilog. Figure 1.29 shows the Active-HDL window at the beginning. Getting Started window is displayed. Here one can select recent projects or create a new one. To continue select create a new workspace.

A wizard is started to create the new workspace. In the firs window the name of the new workspace must be given, see Fig. 1.30. In this example the name given is

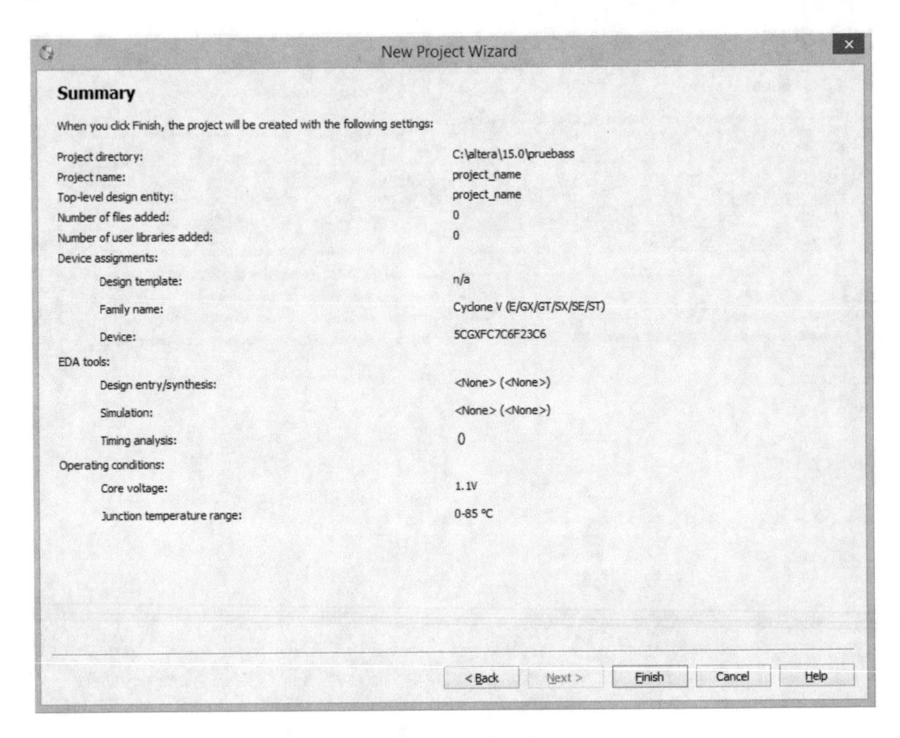


**Fig. 1.27** Quartus II summary (© 2015 Quartus II Altera)

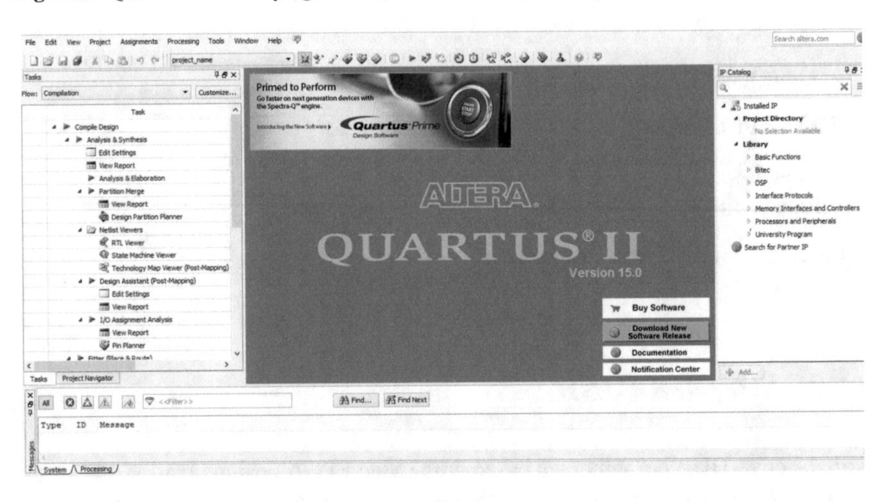


**Fig. 1.28** Quartus II (© 2015 Quartus II Altera)

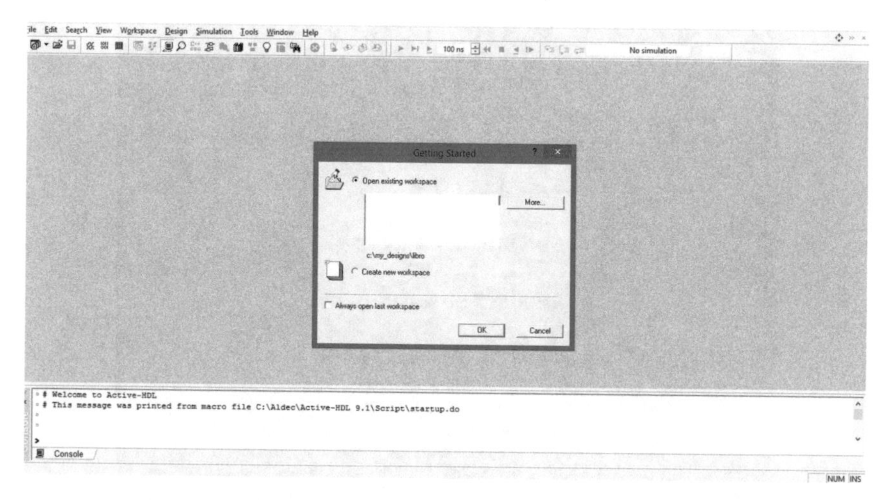


**Fig. 1.29** Active-HDL (© Aldec)

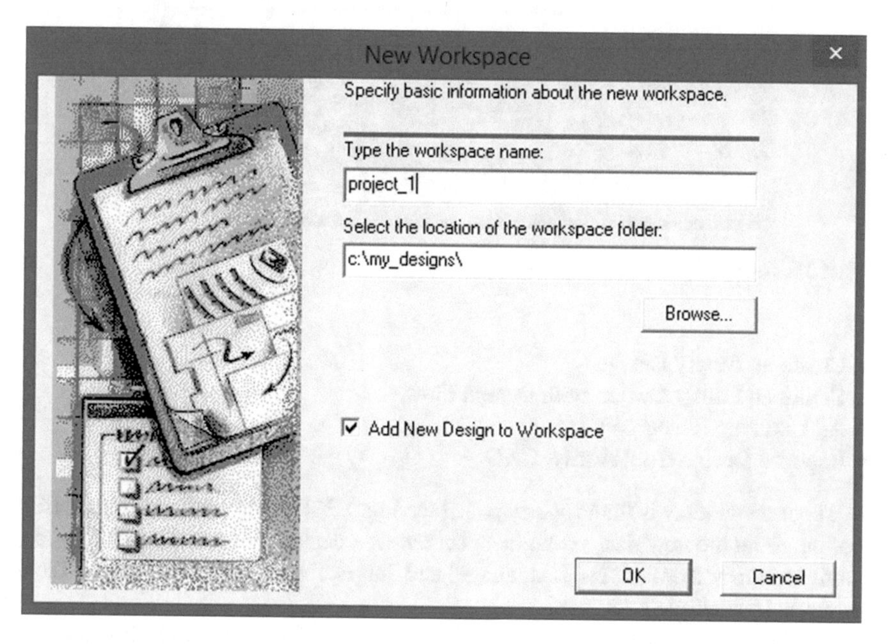


**Fig. 1.30** New workspace window (© Aldec)

project_1 and the location folder is c:\my_desings\. Also the Add New Design to Workspace option is checked. Click Ok to continue.

New Design Wizard window is shown in Fig. 1.31, where four options are presented to select the best option for the design. Then click continue. The options are:

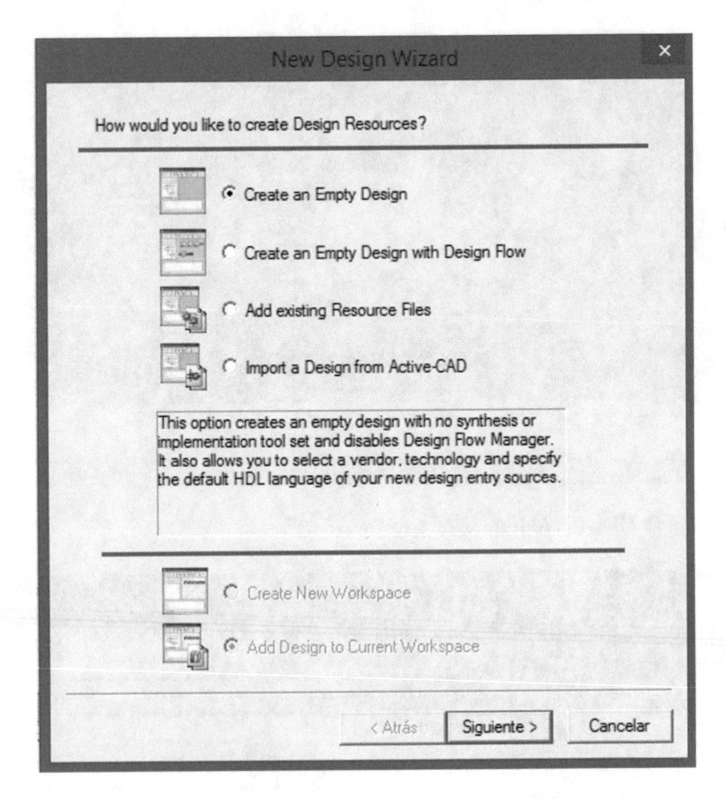


**Fig. 1.31** New Design Wizard window (© Aldec)

- Create an Empty Design
- Create an Empty Design with Design Flow
- Add existing Resource Files
- Import a Design from Active-CAD

The next window is the Property page (see Fig. 1.32), here some additional information about the new design can be specified as a design language (block diagram configuration and default HDL language) and Target Technology (Vendor and Technology). To continue click next.

Into the workspace one or more designs can be declared. Figure 1.33 shows the New Design Name window where the name of the design is given. Also the location of the design folder is given, by default the location is the same of the workspace folder. To better understand the difference of workspace and design imagine a workspace named digital systems, then two designs are declared into the workspace: combinational and sequential. At the same time, each design can have several files for example taking into account the combinational design some files are multiplexors, adders, multipliers, etc.

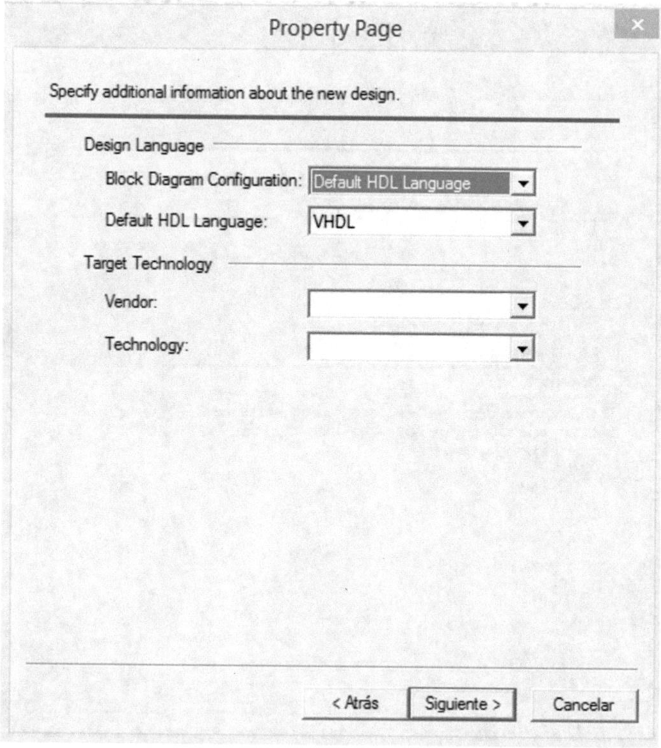

**Fig. 1.32** Property page window (© Aldec)

In the last window of the wizard a summary of the new design is presented. This window is shown in Fig. 1.34. Click next to continue.

Finally, Fig. 1.35 shows the Active-HDL window with the new generated project. It can be seen the design browser subwindow, in which in a tree mode are presented the workspace, the design and the files included into the design. New Files can be added from here.

Figure 1.36 shows an Active-HDL circuit simulation, for this example the AND gate is simulated, the inputs are A (blue signal) and B (green signal), the output is C (red signal). Through the signal stimulation in time the respond of the gate can be tested. A cursor allows to check the value of the inputs and the output in a specific time, in the figure the cursor is around 392 ns, at this moment the signal value are A = 1, B = 1 and C = 1.

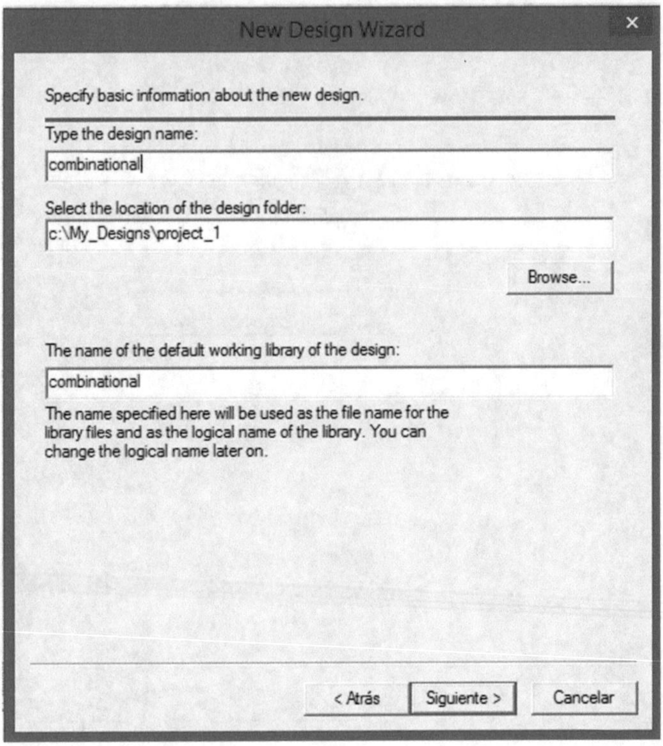

**Fig. 1.33** New Design Name window (© Aldec)

## 1.4 Computer Arithmetic

Our positional numerical system multiplies by powers of 10, starting in 0, each decimal digit to the left of decimal point; and divides by the same powers ($1/10^m = 10^{-m}$) but now starting at 1, the decimal digits to the right of the decimal point. For example, the number 341.12, the three digits to the left of decimal point mean $3 * 100 + 4 * 10 + 1 * 1$ (1 is equal to $10^0$); the three digits if the right mean $1 * 10^{-1} + 2 * 10^{-2}$. The entire number is the sum of both parts.

In the same form works the binary numerical system, but only two digits are available (binary digits, bits, 0 or 1) and the powers use the base 2, instead the base 10 of our decimal system.

The number 1010 means $1 * 2^3 + 0 * 2^2 + 1 * 2^1 + 0 * 2^0 = 8 + 0 + 2 + 0 = 10_d$.

Now we want to represent positive and negative numbers. Using three bits we can represent the decimal number 0–7, as it is shown in Table 1.1 on the first column.

We could attach a digit in front of the number: '0' to present positive numbers, and a '1' to represent the negatives. This representation is called *signed magnitude*, and it is shown in Table 1.1 on column 2. There exist other two possibilities to repre-

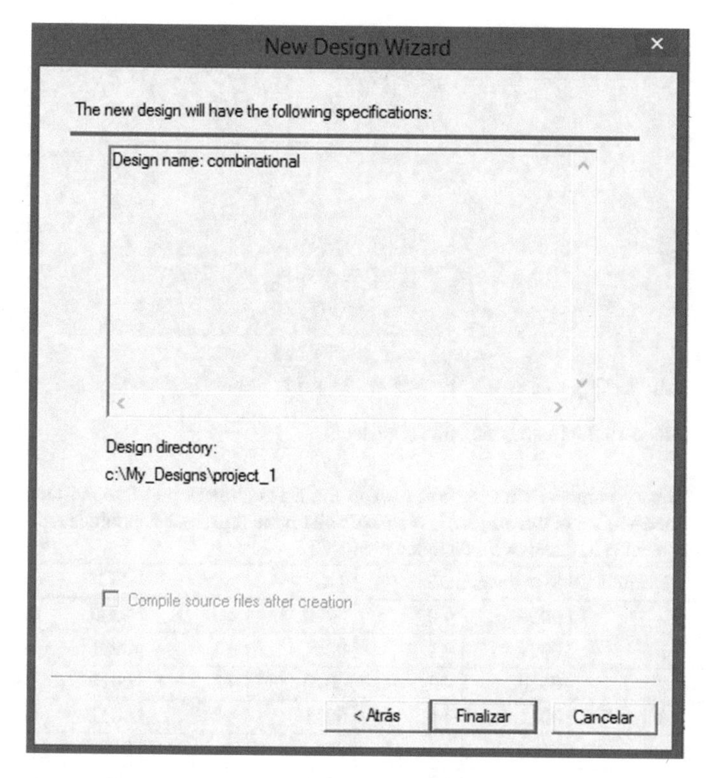


**Fig. 1.34** New design wizard summary (© Aldec)

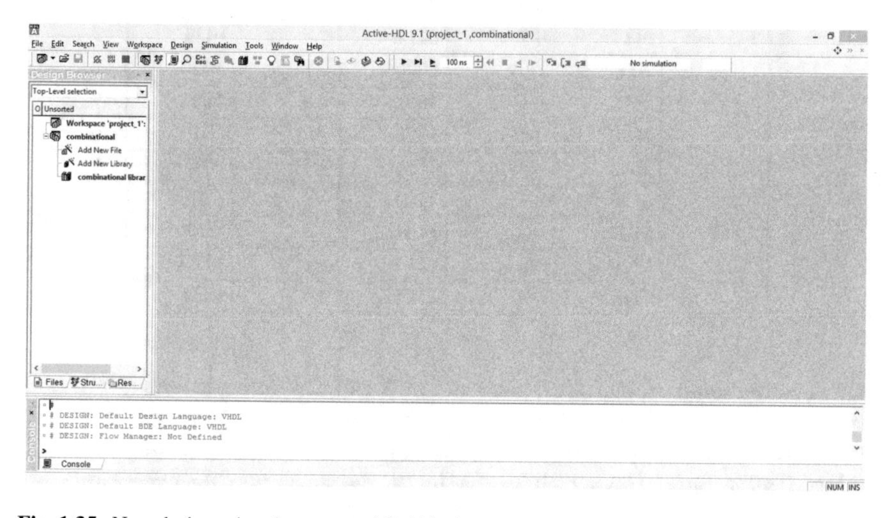


**Fig. 1.35** New design wizard summary (© Aldec)

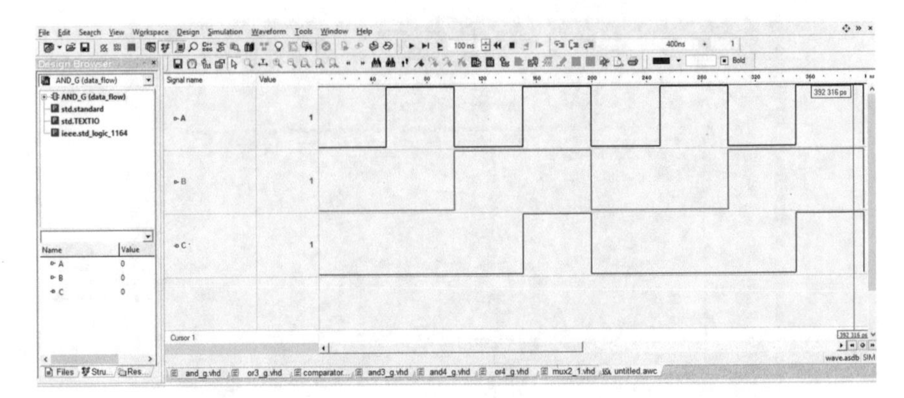


**Fig. 1.36** Active-HDL circuit simulation (© Aldec)

**Table 1.1** Binary numbers with three digits (on the first column), and three different forms to represent signed number of three digits: Use an extra bit for the sign (sign-magnitude representation), one's complement (1'C), and two's complement (2'C)

| Binary | Decimal | Sign-magnitude | | 1'C | | 2'C | |
|---|---|---|---|---|---|---|---|
| 000 | 0 | 0000 | +0 | 0000 | +0 | 0000 | 0 |
| 001 | 1 | 0001 | +1 | 0001 | +1 | 0001 | +1 |
| 010 | 2 | 0010 | +2 | 0010 | +2 | 0010 | +2 |
| 011 | 3 | 0011 | +3 | 0011 | +3 | 0011 | +3 |
| 100 | 4 | 0100 | +4 | 0100 | +4 | 0100 | +4 |
| 101 | 5 | 0101 | +5 | 0101 | +5 | 0101 | +5 |
| 110 | 6 | 0110 | +6 | 0110 | +6 | 0110 | +6 |
| 111 | 7 | 0111 | +7 | 0111 | +7 | 0111 | +7 |
| | | 1000 | −0 | 1000 | −7 | 1000 | −8 |
| | | 1001 | −1 | 1001 | −6 | 1001 | −7 |
| | | 1010 | −2 | 1010 | −5 | 1010 | −6 |
| | | 1011 | −3 | 1011 | −4 | 1011 | −5 |
| | | 1100 | −4 | 1100 | −3 | 1100 | −4 |
| | | 1101 | −5 | 1101 | −2 | 1101 | −3 |
| | | 1110 | −6 | 1110 | −1 | 1110 | −2 |
| | | 1111 | −7 | 1111 | −0 | 1111 | −1 |

sent negative numbers: one's complement (1'C), and two's complement (2'C). Both representations affect only the representation of negative numbers (see Table 1.1). For example, for number 1001, in order to change it to 1'C, it is necessary to negate, or invert, all its bits: 0110, which is $-6_d$. In 2'C representation, the same number is inverted and then added with 1, then 1001 will be $0110 + 1 = 0111 = -7_d$. Note in Table 1.1 that for all negative numbers, invariantly they always start with a 1.

The problem with the signed magnitude and 1'C representations can be seen in Table 1.1: there are two zeros (+0 and −0), a characteristic that can complicate the comparison by zero operations (it is necessary to perform two comparisons instead of a single one). 2'C representation have only one zero. In the rest of this section the 2'C representation will be used.

### 1.4.1 Fixed Point Numbers

A binary number $A(a, b)$ uses $a$ bits to represent its integer part, and $b$ bits to represent its fractional part, then the number will require $a + b + 1$ bits in total (a bit is added because of the sign).

For a binary number $x \in A(a, b)$, the range of numbers that can be presented:

$$-2^a \leq x \leq 2^a - 2^{-b} \tag{1.1}$$

Example: for a number $A(3, 0)$, 3 bits are required to represent the integer part (plus a bit for the sign), and without a fractional part. The range of numbers that can be represented is: $[-8, 7]$.

The range that can be represented for number $A(2, 1)$ is $[-4, 4 - 1/2]$, which is also equal to $[-4, 3.5]$.

**Table 1.2** With 4 bits it can be represented numbers A(3, 0), A(2, 1), A(1, 2), or A(0, 3)

| 2'C | $A(3, 0)$ | $A(2, 1)$ | $A(1, 2)$ | $A(0, 3)$ |
|---|---|---|---|---|
| 0000 | 0 | 0.0 | 0.00 | 0.000 |
| 0001 | 1 | 0.5 | 0.25 | 0.125 |
| 0010 | 2 | 1.0 | 0.50 | 0.250 |
| 0011 | 3 | 1.5 | 0.75 | 0.375 |
| 0100 | 4 | 2.0 | 1.00 | 0.500 |
| 0101 | 5 | 2.5 | 1.25 | 0.625 |
| 0110 | 6 | 3.0 | 1.50 | 0.750 |
| 0111 | 7 | 3.5 | 1.75 | 0.875 |
| 1000 | −8 | −4.0 | −2.00 | −1.000 |
| 1001 | −7 | −3.5 | −1.75 | −0.875 |
| 1010 | −6 | −3.0 | −1.50 | −0.750 |
| 1011 | −5 | −2.5 | −1.25 | −0.625 |
| 1100 | −4 | −2.0 | −1.00 | −0.500 |
| 1101 | −3 | −1.5 | −0.75 | −0.375 |
| 1110 | −2 | −1.0 | −0.50 | −0.250 |
| 1111 | −1 | −0.5 | −0.25 | −0.125 |

This table shows that binary numbers are the same, it just changes the interpretation of the same binary number

As can be seen in Table 1.2, with three bits (plus a bit for the sign), one can represent the numbers A(3, 0), A(2, 1), A(1, 2) or A(0, 3). The binary number does not change, or in other words, the binary point only change the form of interpreting the number.

### 1.4.2 Operations with 2' Complement Numbers

We are going to add two numbers $A(a, b)$, these two numbers have the range $[-2^a, 2^a - 2^{-b}]$. The greatest number that is possible to obtain is by adding two extremes: the two more positive or the two more negative. Adding the two most negative one gives

$$-2^a + (-2^a) = 2(-2^a) = -2^{a+1}.$$

Adding the two more positive:

$$(2^a - 2^{-b}) + (2^a - 2^{-b}) = 2(2^a - 2^{-b}) = 2^{a+1} - 2^{1-b}$$

and $|-2^{a+1}| > |2^{a+1} - 2^{1-b}|$. Then one needs to represent the greatest negative number $-2^{a+1}$.

Therefore, adding two numbers $A(a, b)$ results in a number $A(a+1, b)$. Example: for the number in Table 1.1, using numbers $A(3, 0)$, the greatest number is generated summing $-8 - 8 = -16$. This result needs number $A(4, 0)$, which have range $[-16, 15]$.

For multiplication of two numbers 2'C in $A(a, b)$, we are going to analyze how many bits we need to store the result. The biggest numbers that will be generated are the most positive result generated by multiplying both negative extremes; and the most negative one by multiplying the most positive by the most negative. Then, for the two more negatives:

$$(-2^a)(-2^a) = 2^{2a} \quad \text{(remember, it is positive!)},$$

and the most negative by the most positive:

$$(-2^a)(2^a - 2^{-b}) = -2^{2a} + 2^{a-b},$$

and the multiplication of two more positives is:

$$(2^a - 2^{-b})^2 = 2^{2a} - 2^{-2b}$$

The biggest positive number that is necessary to represent is $2^{2a}$. It is necessary to represent also $-2^{-2b}$, then we need a number $A(2a+1, 2b)$ in order to represent the multiplication of two numbers $A(a, b)$. $A(2a+1, 2b)$ has the range $[-2^{2a+1}, 2^{2a+1} - 2^{2b}]$.

| (a) (4)(5) | (b) (−5)(4) | (c) (−5)(−5) |
|---|---|---|
| 0100 | 11111011 | 11111011 |
| 0101 | 0100 | 11111011 |
| 0100 | 00000000 | 11111011 |
| 0000 | 0000000 | 1111011 |
| 0100 | 111011 | 000000 |
| 0000 | 00000 | 11011 |
| $00010100 = 20_d$ | $11101100 = -20_d$ | 1011 |
| | | 011 |
| | | 11 |
| | | 1 |
| | | $00011001 = 25_d$ |

**Fig. 1.37** Cases of multiplying two numbers, in **a** two positives, **b** a positive and a negative one, and in **c** two negatives

Example: multiplying two numbers $A(3, 0)$ results in:

$$(-8)(-8) = 64 \text{ (positive!)}$$
$$(-8)(7) = -56$$
$$(7)(7) = 49$$

and it is required 7 bits ($2^6 = 64$ plus the sign bit). A number $x \in A(6, 0)$ can represent numbers $-64 \leq x \leq 63$. With a number $y \in A(7, 0)$, one can represent numbers in the range $-128 \leq y \leq 127$.

The most positive is an extreme. We could manage this extreme case as an overflow, and to leave result of multiplying to numbers $A(a, b)$ in $A(2a, 2b)$.

There is no problem in the multiplication of two positive numbers (see Fig. 1.37). If a negative number is involved, then it is necessary to extend the sign bit to the length of the result ($A(2a + 1, 2b)$) and perform the normal procedure, as can be seen in Fig. 1.37.

### 1.4.3 Floating-Point Numbers

Floating-point numbers are defined in the standard IEEE-754 [11], the most common formats are for 32 and 64 bits. A standard floating-point word consists of a sign bit $s$, exponent $e$, and an unsigned normalized mantissa $m$ as arranged as follows:

| $s$ | Exponent $e$ | Unsigned mantissa $m$ |
|---|---|---|

Its algebraic representation is:

$$x = (-1)^s \times 1.m \times 2^{e-b}, \tag{1.2}$$

where $b$ is de bias. This number bias is set based on the number of bits to represent the exponent. If the exponent has $k$ bits then the bias is:

$$b = 2^{k-1} - 1 \tag{1.3}$$

It is preferred to use fixed point operations instead of floating-point ones on FPGAs, because the fixed point notation has higher speed and lower cost, while floating-point notation has higher dynamic range that is quite useful in complicated algorithms [12].

# Chapter 2
# VHDL

## 2.1 A Brief History of VHDL

VHDL is the acronym of *Very High-Speed Integrated Circuit Hardware Description Language*, and it was developed around 1980 at the request of the U.S. Department of Defense. At the beginning, the main goal of VHDL was the electric circuit simulation; however, tools for synthesis and implementation in hardware based on VHDL behavior or structure description files were developed later. With the increasing use of VHDL, the need for standardized was generated. In 1986, the Institute of Electrical and Electronics Engineers (IEEE) standardized the first hardware description language, VHDL, through the 1076 and 1164 standards. VHDL is technology/vendor independent, then VHDL codes are portable and reusable.

## 2.2 VHDL Structure

VHDL is a structured language. Each description of a file has three main blocks:

- Libraries
- Entity
- Architecture

Listing 2.1 shows the main standard libraries for logic and arithmetic descriptions. "Unsigned" and "arith" libraries were developed by Synopsys Inc., they may be under ©.

```
1 library IEEE;
2 use IEEE.std_logic_1164.all;
3 use IEEE.std_logic_unsigned.all;
4 use IEEE.std_logic_arith.all;
5 use IEEE.numeric_std.all;
```

**Listing 2.1** Libraries


E. Tlelo-Cuautle et al., *Engineering Applications of FPGAs*,
DOI 10.1007/978-3-319-34115-6_2

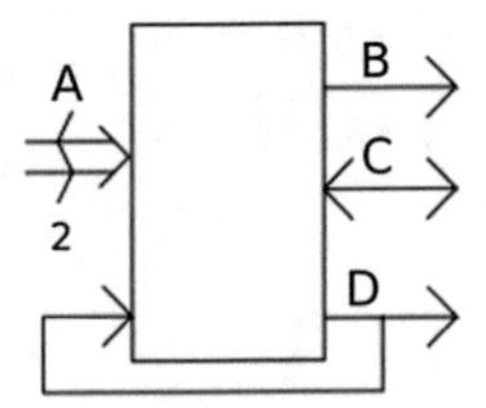


**Fig. 2.1** Black box

Entity can be seen as a black box as shown by Fig. 2.1, where the inputs and outputs must be defined here (see listing 2.2). For example, Fig. 2.1 has four ports: signal A is type **in**, signal B is type **out**, signal C is type **in/out**, and signal D is type **buffer**.

- **in.** Input signal to the entity. Unidirectional
- **out.** Output signal to the entity. Unidirectional
- **in/out.** Input–output signal to the entity. Bidirectional
- **buffer.** Allows internal feedbacks inside the entity. The declared port behavior is as an output.

The data type for each port must be defined. Some of the most used in VHDL are:

- **Bit.** The only values that port allows are 0 or 1.
- **Boolean.** Take the values true or false.
- **Integer.** This type cover all integer values.
- **std_logic.** This data type allows nine values
  - **U** Unitialized
  - **X** Unknown
  - **0** Low
  - **1** High
  - **Z** High impedance
  - **W** Weak unknown
  - **L** Weak low
  - **H** Weak high
  - **'-'** Don't care
- **bit_vector.** A vector of bits.
- **std_logic_vector.** A vector of bits of type std_logic.

```
1 entity name_of_entity is
2   port(
3   port_name: port_mode signal_type;
4   port_name: port_mode signal_type;
5   .....
6   );
7 end [entity] [name_of_entity];
```

**Listing 2.2** Entity declaration

Listing 2.3 shows the entity description for the black box of Fig. 2.1.

```
1 entity black_bok is
2   port(
3   A : in   std_logic_vector(1 downto 0);
4   B : out   std_logic;
5   C : inout std_logic;
6   D : buffer std_logic
7   );
8 end black_box;
```

**Listing 2.3** Entity black box

Architecture contains a description of how the circuit should function, from which the actual circuit is inferred. A syntax for an architecture description is shown in listing 2.4.

```
1 architecture architecture_name of entity_name is
2 [architecture_declarative_part]
3 begin
4 architecture_statements_part
5 end [architecture] [architecture_name];
```

**Listing 2.4** Architecture syntax

Listing 2.5 shows an example of an architecture description for an AND gate. A complete description of the AND gate including libraries and entity is shown in listing 2.6. You may check the next related books [13, 14, 15, 16].

```
1 architecture example of AND_G is
2 begin
3   C <= A AND B;
4   — This is a comment
5   — C is an output
6   — A, B are inputs
7 end architecture example;
```

**Listing 2.5** Architecture of AND gate

```
1 library IEEE;
2 use IEEE.std_logic_1164.all;
3
4 entity AND_G is
5   port(
6   A : in   std_logic;
7   B : in   std_logic;
8   C : out std_logic
9   );
10 end AND_G;
11
12 architecture example of AND_G is
13 begin
14   C <= A AND B;
15 end architecture example;
```

**Listing 2.6** AND gate description

## 2.3 Levels of Abstraction

VHDL allows different styles for architecture description, they can be classified as:

- Behavioral description
- Structural description
- Data flow description

### 2.3.1 *Behavioral Description*

Behavioral description reflects the system function, how the system works without taking care about the elements that compose it. It is just a relation between inputs and outputs. A process structure is present in a combinational description. For example, listing 2.7 shows a behavioral description for a XOR gate. For this example it is considered that (Fig. 2.2 and Table 2.1):

$$\text{if A} = \text{B then C} = 0$$

$$\text{if A} \neq \text{B then C} = 1$$

```
1  library IEEE;
2  use IEEE.std_logic_1164.all;
3
4  entity XOR_G is
5    port(
6    A : in  std_logic;
7    B : in  std_logic;
8    C : out std_logic
9    );
10 end XOR_G;
11
12 architecture behavioral of XOR_G is
13 begin
14   process(A,B)
15   begin
16     if A = B then
17       C <= '0';
18     else
19       C <= '1';
20     end if;
21   end process;
22 end architecture behavioral;
```

**Listing 2.7** XOR gate behavioral description

Another example is shown in listing 2.8. It shows the behavioral description for the AND gate considering that (Fig. 2.3 and Table 2.2):

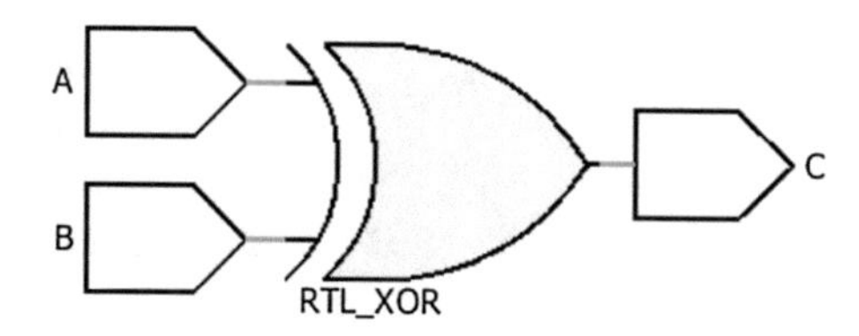


**Fig. 2.2** RTL XOR

**Table 2.1** XOR true table

| A | B | C |
|---|---|---|
| 0 | 0 | 0 |
| 0 | 1 | 1 |
| 1 | 0 | 1 |
| 1 | 1 | 0 |

$$\text{if } A = 1 \text{ and } B = 1 \text{ then } C = 1$$

$$\text{other case } C = 0$$

```
1  library IEEE;
2  use IEEE.std_logic_1164.all;
3
4  entity AND_G is
5    port(
6    A : in  std_logic;
7    B : in  std_logic;
8    C : out std_logic
9    );
10 end AND_G;
11
12 architecture behavioral of AND_G is
13 begin
14   process(A,B)
15   begin
16     if A = '1' and B = '1' then
17       C <= '1';
18     else
19       C <= '0';
20     end if;
21   end process;
22 end architecture behavioral;
```

**Listing 2.8** AND gate behavioral description

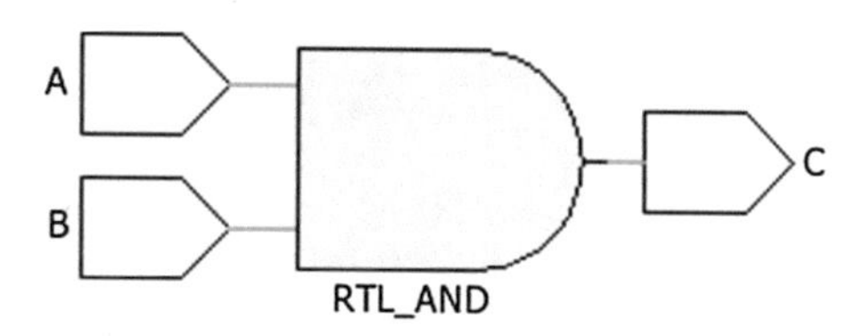


**Fig. 2.3** RTL AND

**Table 2.2** AND true table

| A | B | C |
|---|---|---|
| 0 | 0 | 0 |
| 0 | 1 | 0 |
| 1 | 0 | 0 |
| 1 | 1 | 1 |

**Table 2.3** 2-bit comparator true table

| A | B | G | E | L |
|---|---|---|---|---|
| 00 | 00 | 0 | 1 | 0 |
| 00 | 01 | 0 | 0 | 1 |
| 00 | 10 | 0 | 0 | 1 |
| 00 | 11 | 0 | 0 | 1 |
| 01 | 00 | 1 | 0 | 0 |
| 01 | 01 | 0 | 1 | 0 |
| 01 | 10 | 0 | 0 | 1 |
| 01 | 11 | 0 | 0 | 1 |
| 10 | 00 | 1 | 0 | 0 |
| 10 | 01 | 1 | 0 | 0 |
| 10 | 10 | 0 | 1 | 0 |
| 10 | 11 | 0 | 0 | 1 |
| 11 | 00 | 1 | 0 | 0 |
| 11 | 01 | 1 | 0 | 0 |
| 11 | 10 | 1 | 0 | 0 |
| 11 | 11 | 0 | 1 | 0 |

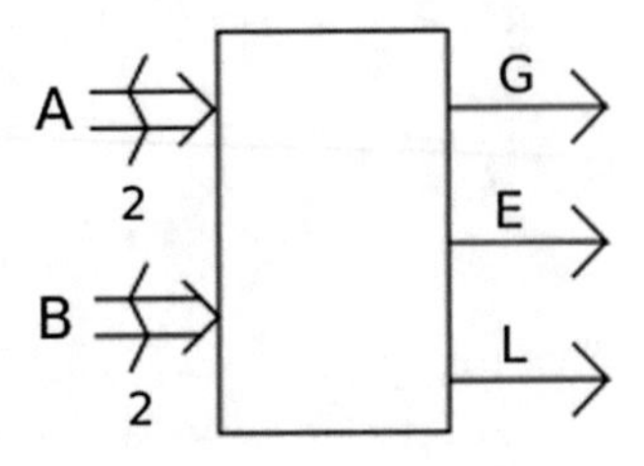


**Fig. 2.4** 2-bit comparator

Listing 2.9 shows the behavioral description of a 2-bit comparator (Table 2.3). Figure 2.4 shows the inputs and outputs of the 2-bit comparator. For the behavioral description it is considered that:

$$\text{if } A = 1 \text{ and } B = 1 \text{ then } C = 1$$

$$\text{other case } C = 0$$

```
1  library IEEE;
2  use IEEE.std_logic_1164.all;
3
4  entity comparator_2bits is
5    port(
6    A : in  std_logic_vector(1 downto 0);
7    B : in  std_logic_vector(1 downto 0);
8    G : out std_logic;
9    E : out std_logic;
10   L : out std_logic
11   );
12 end comparator_2bits;
13
14 architecture behavioral of comparator_2bits is
15 begin
16   combinational: process(A,B)
17   begin
18     if A > B then
19       G<= '1';
20     else
21       G<= '0';
22     end if;
23
24     if A = B then
25       E<= '1'
26     else
27       E<= '0';
28     end if;
29
30     if A < B then
31       L<= '1';
32     else
33       L<= '0';
34     end if;
35   end process combinational;
36
37 end architecture behavioral;
```

**Listing 2.9** 2-bit comparator behavioral description

### *2.3.2 Data Flow Description*

Data flow description designates the way how data can be transferred from one signal to another without using sequential statements. The data flow descriptions are concurrent; these kinds of descriptions allow to define the flow that data take from one module to another. An example of data flow description is shown in listing 2.10 (Table 2.4, Fig. 2.5):

$$\text{if } A = 1 \text{ and } B = 1 \text{then } C = 0$$

**Table 2.4** NAND true table

| A | B | C |
|---|---|---|
| 0 | 0 | 1 |
| 0 | 1 | 1 |
| 1 | 0 | 1 |
| 1 | 1 | 0 |

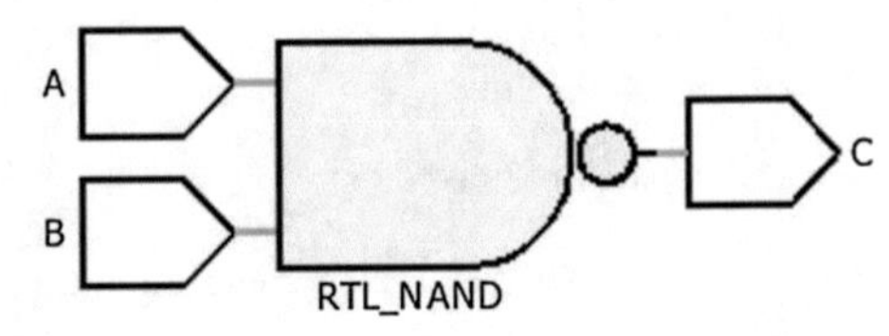


**Fig. 2.5** RTL NAND

**Table 2.5** OR true table

| A | B | C |
|---|---|---|
| 0 | 0 | 0 |
| 0 | 1 | 1 |
| 1 | 0 | 1 |
| 1 | 1 | 1 |

$$\text{other case } C = 1$$

```
1  library IEEE;
2  use IEEE.std_logic_1164.all;
3
4  entity NAND_G is
5    port(
6    A : in  std_logic;
7    B : in  std_logic;
8    C : out std_logic
9    );
10 end NAND_G;
11
12 architecture Data_flow of NAND_G is
13 begin
14
15   C <= '0' when (A = '1' and B = '1') else '1';
16
17 end architecture Data_flow;
```

**Listing 2.10** NAND gate data flow description

Another example of data flow description is shown in listing 2.11. In this case, the data flow description for the OR gate considers that (Table 2.5, Fig. 2.6):

$$\text{if } A = 0 \text{ and } B = 0 \text{ then } C = 0$$

$$\text{other case } C = 1$$

**Fig. 2.6** RTL OR

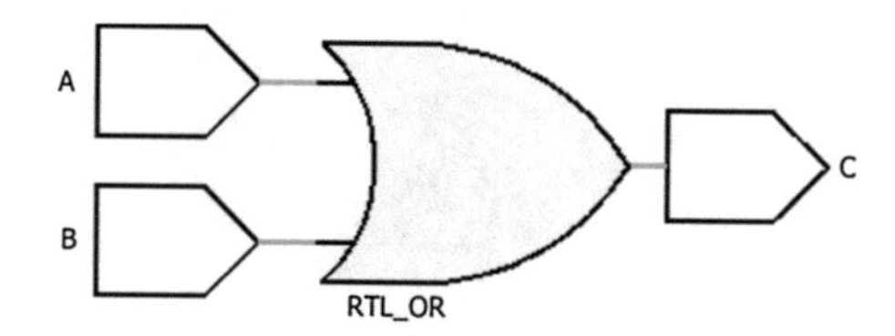


```
library IEEE;
use IEEE.std_logic_1164.all;

entity OR_G is
  port(
  A : in  std_logic;
  B : in  std_logic;
  C : out std_logic
  );
end OR_G;

architecture Data_flow of OR_G is
begin

  C <= '0' when (A = '0' and B = '0') else '1';

end architecture Data_flow;
```

**Listing 2.11** OR gate data flow description

Listing 2.9 shows the data flow description of a 2-bit comparator. Figure 2.4 shows the inputs and output of the 2-bit comparator and Table 2.3 its True Table.

```
library IEEE;
use IEEE.std_logic_1164.all;

entity comparator_2bits is
  port(
  A : in  std_logic_vector(1 downto 0);
  B : in  std_logic_vector(1 downto 0);
  G : out std_logic;
  E : out std_logic;
  L : out std_logic
  );
end comparator_2bits;

architecture data_flow of comparator_2bits is
begin

  G <= '1' when A > B else '0';
  E <= '1' when A = B else '0';
  L <= '1' when A < B else '0';

end architecture data_flow;
```

**Listing 2.12** 2-bit comparator data flow description

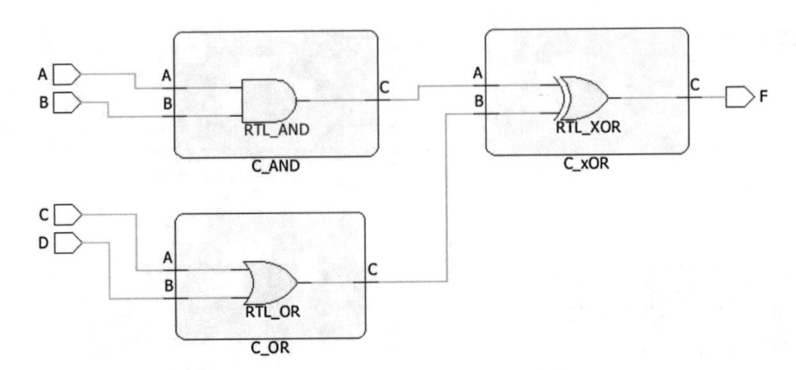


**Fig. 2.7** RTL EXAMPLE

## *2.3.3 Structural Description*

Structural description is based on established logic models (gates, adders, counters, etc.), which are called as components and they are interconnected in a netlist. Structural description has a hierarchy, it is necessary to reduce the design in small modules (components). These components will be called into another module of more hierarchy. This reduction allows a practical analysis of small modules and it is a simple form to describe.

Figure 2.7 shows an example of structural description, in this example are used the AND, OR, XOR gates described above. Entity "example" is the top level design. Listing 2.13 shows the structural description for the example.

```
1  library ieee;
2  use ieee.std_logic_1164.all;
3
4  entity example is
5      port (
6      A : in  std_logic;
7      B : in  std_logic;
8      C : in  std_logic;
9      D : in  std_logic;
10     F : out std_logic
11         );
12 end example;
13
14 architecture structural of example is
15 component AND_G
16     port(
17     A : in    std_logic;
18     B : in    std_logic;
19     C : out   std_logic
20     );
21 end component;
22 component OR_G
23     port(
```

```
24      A : in    std_logic;
25      B : in    std_logic;
26      C : out   std_logic
27      );
28  end component;
29  component XOR_G
30      port(
31      A : in    std_logic;
32      B : in    std_logic;
33      C : out   std_logic
34      );
35  end component;
36
37  signal SI0, SI1, SI2, SI3, SI4 : std_logic;
38
39  begin
40      DUT1  : AND_G  port map(A,B,SI0);
41      DUT2  : XOR_G  port map(SI0,SI1,F);
42      DUT3  : OR_G   port map(C,D,SI1);
43
44  end structural;
```

**Listing 2.13** Structural Description Example

Listing 2.14 is the structural description of the 2-bit comparator, its RTL was divided into three sections. The first one corresponds to G signal, when A is greater than B, the RTL is shown in Fig. 2.8. The second one shows the RTL for E signal, when A is equal to B, in this case Fig. 2.9 shows its RTL. Finally, the RTL for signal L is shown in Fig. 2.10, when A is lower that B. This example for the structural description of a 2-bit comparator, shows different levels of abstraction, beginning with gates, their interconnections into a more complex gates (for example the OR4_G is an OR with four inputs), the description of a logic function (G, E, L) and finally a combinational circuit (comparator).

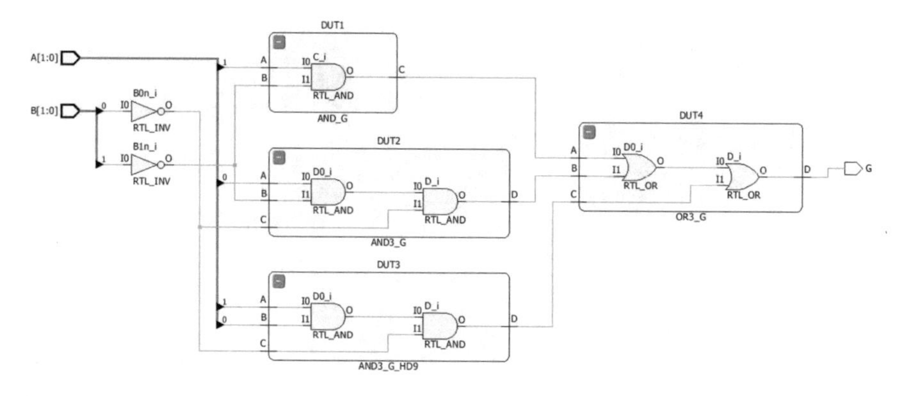


**Fig. 2.8** RTL signal G

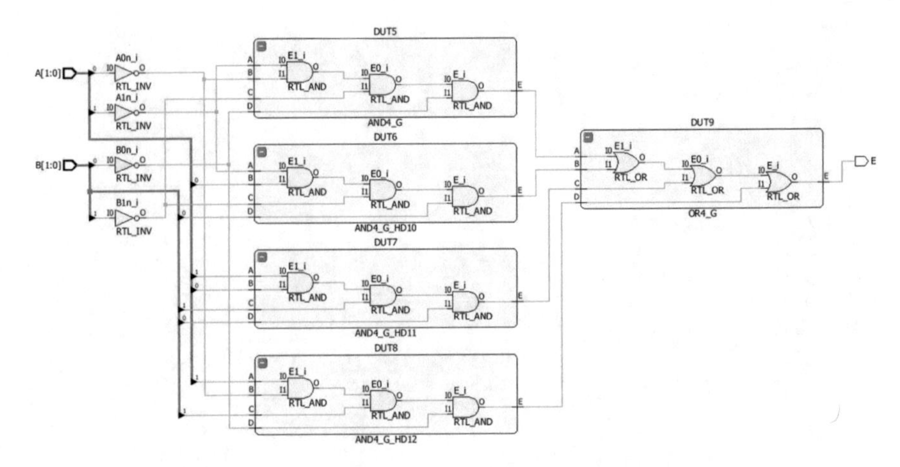


Fig. 2.9 RTL signal E

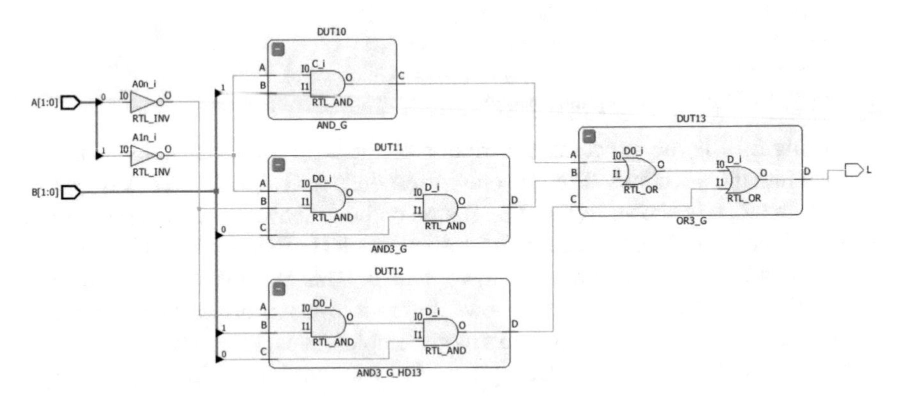


Fig. 2.10 RTL signal L

```
1  library ieee;
2  use ieee.std_logic_1164.all;
3
4  entity comparator_2bits is
5    port (
6    A : in  std_logic_vector(1 downto 0 );
7    B : in  std_logic_vector(1 downto 0 );
8    G : out std_logic;    — A > B
9    E : out std_logic;    — A = B
10   L : out std_logic    — L < B
11     );
12 end comparator_2bits;
13
14 architecture structural of comparator_2bits is
15
16 component AND_G
17   port(
```

```
18    A : in  std_logic;
19    B : in  std_logic;
20    C : out std_logic
21    );
22  end component;
23
24  component OR3_G
25    port(
26    A : in  std_logic;
27    B : in  std_logic;
28    C : in  std_logic;
29    D : out std_logic
30    );
31  end component;
32
33  component AND4_G
34    port(
35    A : in  std_logic;
36    B : in  std_logic;
37    C : in  std_logic;
38    D : in  std_logic;
39    E : out std_logic
40    );
41  end component;
42
43  component AND3_G
44    port(
45    A : in  std_logic;
46    B : in  std_logic;
47    C : in  std_logic;
48    D : out std_logic
49    );
50  end component;
51
52  component OR4_G
53    port(
54    A : in  std_logic;
55    B : in  std_logic;
56    C : in  std_logic;
57    D : in  std_logic;
58    E : out std_logic
59    );
60  end component;
61
62  signal A1n,A0n, B0n,B1n : std_logic;
63  signal S1,S2,S3,S4,S5,S6,S7,S8,S9,S10 : std_logic;
64
65  begin
66    B0n <= not B(0);
67    B1n <= not B(1);
68    A0n <= not A(0);
69    A1n <= not A(1);
70
```

```
71  ——— G————
72  DUT1: AND_G  port map(A(1),B1n,S1);
73  DUT2: AND3_G port map(A(0),B1n,B0n,S2);
74  DUT3: AND3_G port map(A(1),A(0),B0n,S3);
75  DUT4: OR3_G  port map(S1,S2,S3,G);
76
77  ——— E————
78  DUT5: AND4_G port map(A1n,A0n,B1n,B0n,S4);
79  DUT6: AND4_G port map(A1n,A(0),B1n,B(0),S5);
80  DUT7: AND4_G port map(A(1),A(0),B(1),B(0),S6);
81  DUT8: AND4_G port map(A(1),A0n,B(1),B0n,S7);
82  DUT9: OR4_G  port map(S4,S5,S6,S7,E);
83
84  ——— L————
85  DUT10: AND_G  port map(A1n,B(1),S8);
86  DUT11: AND3_G port map(A1n,A0n,B(0),S9);
87  DUT12: AND3_G port map(A0n,B(1),B(0),S10);
88  DUT13: OR3_G  port map(S8,S9,S10,L);
89
90 end structural;
```

**Listing 2.14** 2-bit Comparator structural description

## 2.4 Modules Description Examples

In this section, a descriptions and simulation of some common circuits are given. For example, the blocks: multiplexor, adder, decoder, flip_flop, registers, and counters.

### *2.4.1 Combinational Circuits*

Some gates were described above, so the first example is a simple multiplexer 2 to 1. Mux2_1 RTL is shown in Fig. 2.11 and its description in listing 2.15. Its simulation usign Active-HDL is presented in Fig. 2.12

```
1  library ieee;
2  use ieee.std_logic_1164.all;
3
4  entity mux2_1 is
5    port (
6    I0 : in  std_logic;
7    I1 : in  std_logic;
8    S  : in  std_logic;
9    Y  : out std_logic
10     );
11 end mux2_1;
12
13 architecture data_flow of mux2_1 is
```

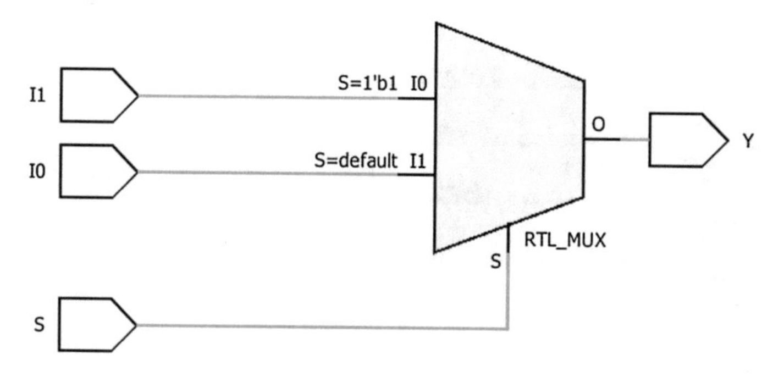


**Fig. 2.11** Mux2_1

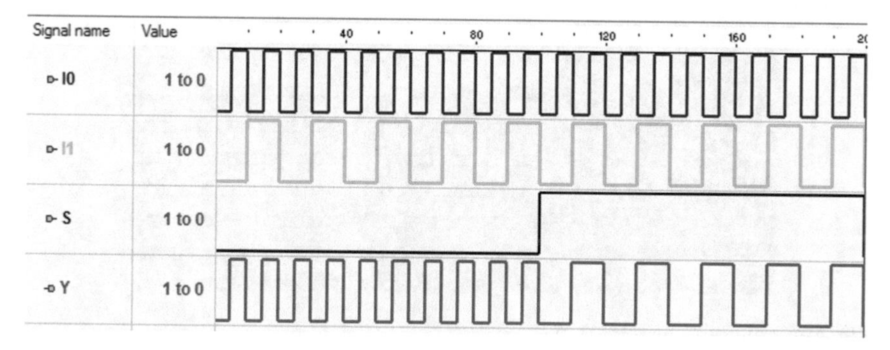


**Fig. 2.12** Simulation of Mux2_1

```
14 begin
15    Y <= I0 when S = '0' else I1;
16 end data_flow;
```

**Listing 2.15** Mux2_1 description

Figure 2.13 presents a multiplexor 4 to 1, but in this case each input is a vector of "n-bit" except the input S which has 2-bit width and it does not depend on the generic n. To declare n the keyword generic is used as is shown in listing 2.16, in line 5. n is the integer type and its default value is four. By using the generic keyword, the value of the vector width can be modified when the multiplexer is used as a component. The description used the with/select structure, for the last case ("11") the keyword others is applied, others included all the combination described for std_logic signals (see Sect. 2.2). The simulation of the mux4_1_n is shown in Fig. 2.14.

```
1 library IEEE;
2 use IEEE.STD_LOGIC_1164.ALL;
3
4 entity mux4_1_n is
5    generic(n : integer := 4);
6    Port ( I0 : in  STD_LOGIC_VECTOR (n−1 downto 0);
7           I1 : in  STD_LOGIC_VECTOR (n−1 downto 0);
```

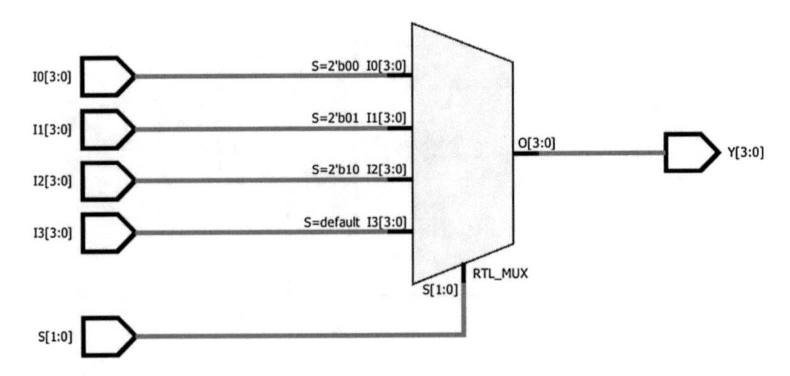


**Fig. 2.13** Mux4_1_n

| Signal name | Value | 0–100 | 100–200 | 200–300 | 300–400 |
|---|---|---|---|---|---|
| ⊞ ▷- I0 | 0 | 0 | | | |
| ⊞ ▷- I1 | 2 | 2 | | | |
| ⊞ ▷- I2 | 7 | 7 | | | |
| ⊞ ▷- I3 | A | A | | | |
| ⊞ ▷- S | 0 | 0 | 1 | 2 | 3 |
| ⊞ -▷ Y | 0 | 0 | 2 | 7 | A |

**Fig. 2.14** Simulation of Mux4_1_n

```
8        I2 : in  STD_LOGIC_VECTOR (n−1 downto 0);
9        I3 : in  STD_LOGIC_VECTOR (n−1 downto 0);
10       S  : in  STD_LOGIC_VECTOR (1   downto 0);
11       Y  : out STD_LOGIC_VECTOR (n−1 downto 0)
12       );
13  end mux4_1_n;
14
15  architecture data_flow of mux4_1_n is
16
17  begin
18    with S select
19      Y <= I0 when "00",
20         I1 when "01",
21         I2 when "10",
22         I3 when others;
23
24  end data_flow;
```

**Listing 2.16** Mux4_1_n description

An example of hexadecimal to 7 segments decoder is shown below. Figure 2.15 shows one input vector of 4 bits and one output vector of 7 bits, for this example the description is behavioral. Simulation for hexadecimal to 7 segments decoder is presented in Fig. 2.16. Please check that segment “a” corresponds to bit seg(7), “b”

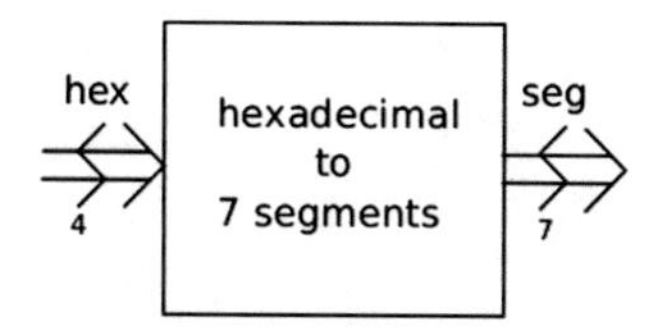


**Fig. 2.15** Hexadecimal to 7 segments decoder

**Fig. 2.16** Simulation hexadecimal to 7 segments decoder

to seg(6), “c” to seg(5), “d” to seg(4), “e” to seg(3), “f” to seg(2), and “g” to seg(1), for this description the signal seg does not have a bit seg(0) declared.

```
1  library ieee;
2  use ieee.std_logic_1164.all;
3
4  entity hex_7seg is
5    port (
6    hex : in   std_logic_vector(3 downto 0);
7    seg    : out std_logic_vector(7 downto 1)
8        );
9  end hex_7seg;
10
11 architecture behavioral of hex_7seg is
12 begin
13   process(hex)
14   begin
15     case hex is             — abcdefg
16       when x"0" => Seg <= "1111110";
17       when x"1" => Seg <= "0110000";
18       when x"2" => Seg <= "1101101";
19       when x"3" => Seg <= "1111001";
20       when x"4" => Seg <= "0110011";
21       when x"5" => Seg <= "1011011";
22       when x"6" => Seg <= "1011111";
23       when x"7" => Seg <= "1110000";
24       when x"8" => Seg <= "1111111";
25       when x"9" => Seg <= "1111011";
26       when x"A" => Seg <= "1110111";
27       when x"b" => Seg <= "0011111";
28       when x"C" => Seg <= "1001110";
29       when x"d" => Seg <= "0111101";
```

```
30        when x"E" => Seg <= "1001111";
31        when others => Seg <= "1000111"; — F
32      end case;
33    end process;
34
35  end behavioral;
```

**Listing 2.17** Hexadecimal to 7 segments decoder description

Figure 2.17 shows an RTL for a complete adder of 4-bit. The adder has three inputs, two vectors of 4 bits (A and B) and one signal of 1 bit (Cin), and two outputs, one signal of one bit (Cou) and one vector of 4 bits (Sum). A and B are the numbers to be added, Cin is the input carry, Cou is the output carry and Sum is the result of the sum. Listing 2.18 shows a generic adder description, it can be seen in line 6 a generic integer and its default value set to 4. Three internal signals of unsigned type are used for data conversion and to store the internal sum (C, Ai, Bi). These conversions are shown in lines 22 and 23 and the sum in line 26. Finally, the result is converted to std_logic type in lines 29 and 30. The adder simulation is shown in Fig. 2.18.

```
1  library ieee;
2  use ieee.std_logic_1164.all;
3  use ieee.numeric_std.all;
4
```

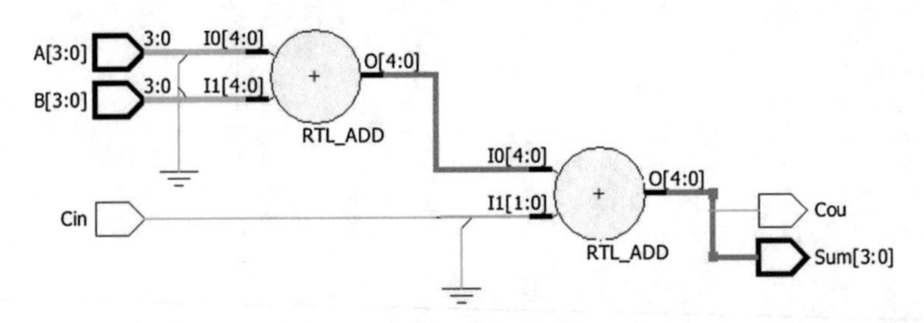


**Fig. 2.17** RTL adder

**Fig. 2.18** Simulation adder

```
5  entity adder_n is
6    generic(n : integer := 4);
7    port (
8    A   : in  std_logic_vector(n-1 downto 0);
9    B   : in  std_logic_vector(n-1 downto 0);
10   Cin : in  std_logic;
11   Sum : out std_logic_vector(n-1 downto 0);
12   Cou : out std_logic
13     );
14 end adder_n;
15
16 architecture behavioral of adder_n is
17 signal C : unsigned(n downto 0);
18 signal Ai,Bi : unsigned(n-1 downto 0);
19 begin
20
21   -- data conversion to unsigned
22   Ai <= unsigned(A);
23   Bi <= unsigned(B);
24
25   -- adder
26   C <= ('0' & Ai) + ('0' & Bi) + ('0' & Cin);
27
28   -- data conversion to std_logic
29   Sum <= std_logic_vector(C(n-1 downto 0));
30   Cou <= std_logic(C(n));
31
32 end behavioral;
```

**Listing 2.18** Adder description

## 2.4.2 *Sequential Circuits*

A basic element in sequential logic is the flip_flop D, its RTL view is shown in Fig. 2.19. The inputs for the flip_flop are: asynchronous reset (RST), clock (CLK), and datum (D), the only output signal is Q. The simulation of the flip_flop is presented in Fig. 2.20, here one can see how Q takes the value of D when the clock transition is positive and holds this value until a new clock transition is presented. The behavioral description of the flip_flop is presented in listing 2.19

```
1  library ieee;
2  use ieee.std_logic_1164.all;
3
4  entity flip_flop_d is
5    port (
6    RST : in  std_logic;
7    CLK : in  std_logic;
8    D   : in  std_logic;
9    Q   : out std_logic
10     );
```

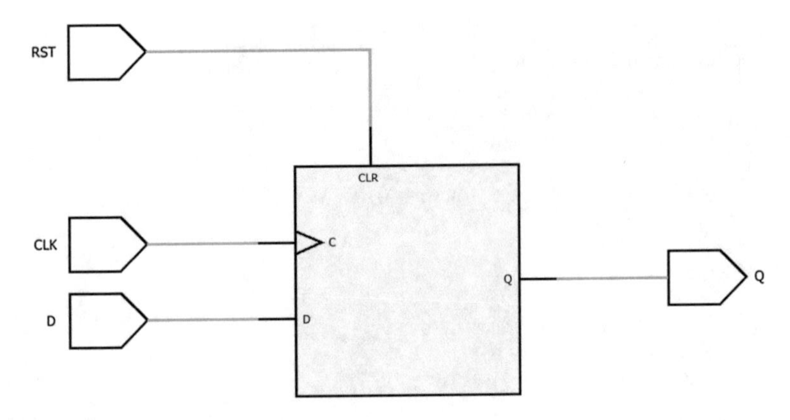


**Fig. 2.19** RTL Flip_Flop D

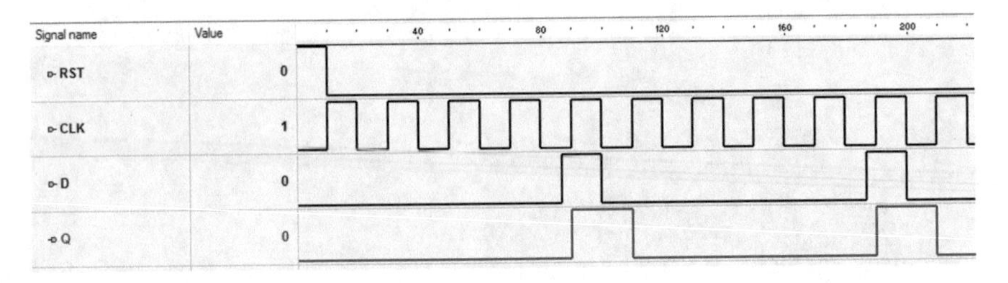


**Fig. 2.20** Simulation Flip_Flop D

```
11 end flip_flop_d;
12
13 architecture behavioral of flip_flop_d is
14 begin
15   process(RST,CLK)
16   begin
17     if RST = '1' then
18       Q<= '0';
19     elsif rising_edge(CLK) then
20       Q<= D;
21     end if;
22   end process;
23
24 end behavioral;
```

**Listing 2.19** Flip_Flop D description

Figure 2.21 shows a RTLRTL of a parallel-parallel enable register of four bits, each bit is stored in a flip_flop. Its inputs are: asynchronous reset (RST), clock (CLK), enable (E), and data (D), the output signal is a vector Q. The simulation of the register can be seen in Fig. 2.22. In the simulation is noted the register behavior, apart from the clock, enable signal must be activated to load D, until E is active, the output Q

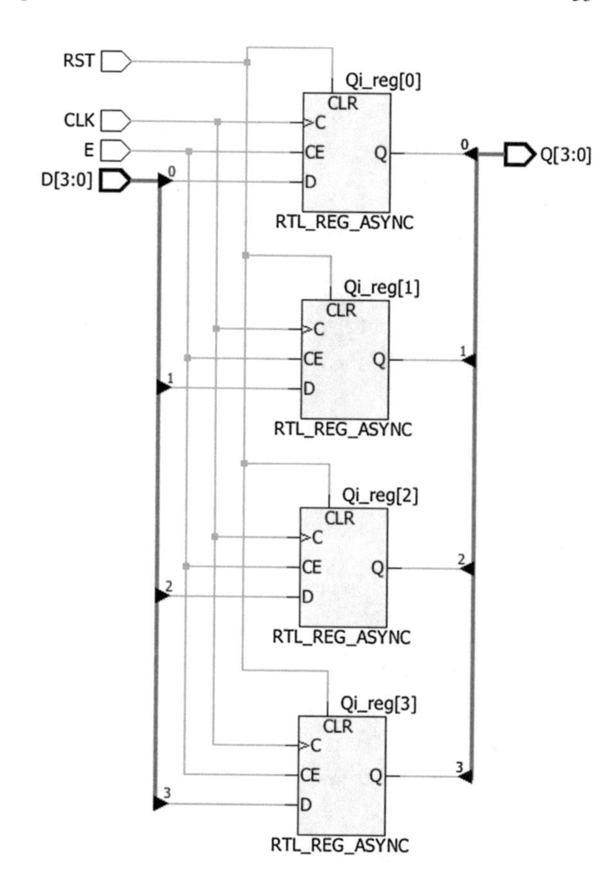


**Fig. 2.21** RTL parallel–parallel enable register

takes the value of D (in each positive clock transition), when E is not active Q holds the last data.

Listing 2.20 is the description of the register, with a generic width. In this example, it was necessary an internal signal Qi. Line 18 assigns Qi to the output Q. The enable is described in lines 25–29, when E = 1 the register load the data D, when E = 0 it holds the previous values.

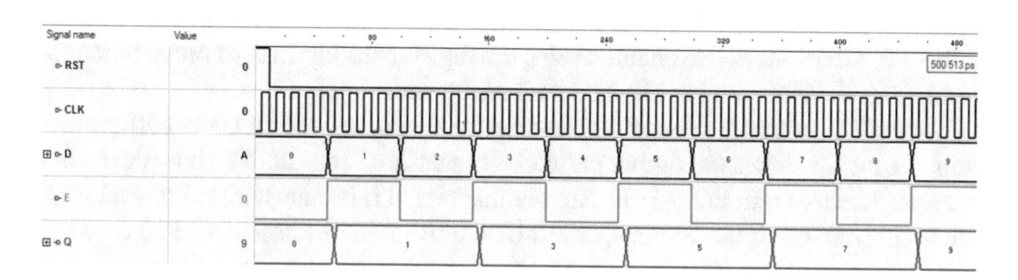


**Fig. 2.22** Simulation parallel–parallel enable register

```
library ieee;
use ieee.std_logic_1164.all;

entity register_epp is
  generic(n : integer := 4);
  port(
  RST : in  std_logic;
  CLK : in  std_logic;
  D   : in  std_logic_vector(n-1 downto 0);
  E   : in  std_logic;
  Q   : out std_logic_vector(n-1 downto 0)
  );
end register_epp;

architecture behavioral of register_epp is
signal Qi : std_logic_vector(n-1 downto 0);
begin
  Q<= Qi;

   process(RST,CLK)
   begin
     if RST = '1' then
       Qi <= (others => '0');
     elsif rising_edge(CLK) then
       if E = '1' then
         Qi <= D;
       else
         Qi <= Qi;
       end if;
     end if;
   end process;

end behavioral;
```

**Listing 2.20** Parallel–parallel enable register description

The next example is a left-shift register with a parallel output, the RTL view is shown in Fig. 2.23, to simplify the RTL, common signals (asynchronous reset RST, clock CLK, and enable E) were removed. One can see how the data flow from flip_flop Qi(0) to Q(1) ... and so on, and the output signal takes the value in a parallel way.

Simulation of the left-shift register with parallel output is shown in Fig. 2.24. L was fixed with a value of one. The register loads this value when the enable (E) is equal to 1 and the clock (CLK) is in a positive transition. Previous values are moved to the left, after four active enable cycles the register is fully load of ones. When the enable is E = '0' the register holds its present value.

Listing 2.21 shows the behavioral description for the left-shift register with parallel output. Line 18 shows the output parallel assignation. Lines 25–29 show the enable and shift functions. In line 26 one can see that signal L is concatenated to vector Qi, due to this one bit of the vector Qi must be removed, in this case the MSB (n–1).

```
library ieee;
use ieee.std_logic_1164.all;
```

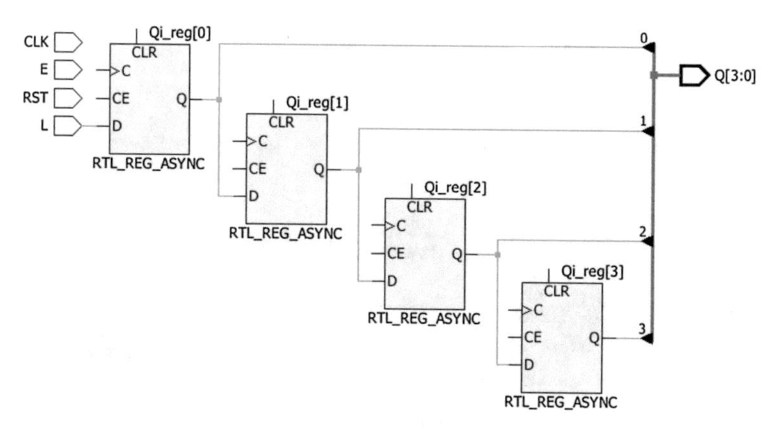


**Fig. 2.23** RTL *left*-shift register parallel output

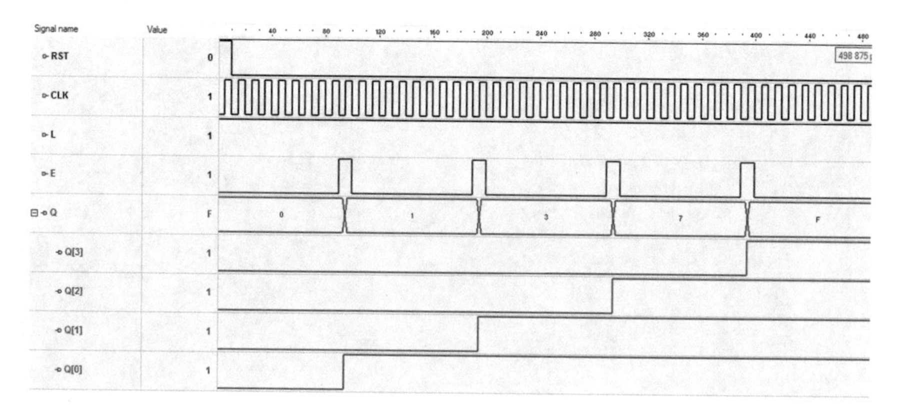


**Fig. 2.24** Simulation *left*-shift register parallel output

```
3
4  entity reg_shift is
5    generic(n : integer := 4);
6    port (
7    RST : in  std_logic;
8    CLK : in  std_logic;
9    L   : in  std_logic;
10   E   : in  std_logic;
11   Q   : out std_logic_vector(n-1 downto 0)
12       );
13 end reg_shift;
14
15 architecture behavioral of reg_shift is
16 signal Qi : std_logic_vector(n-1 downto 0);
17 begin
18   Q <= Qi;
19
```

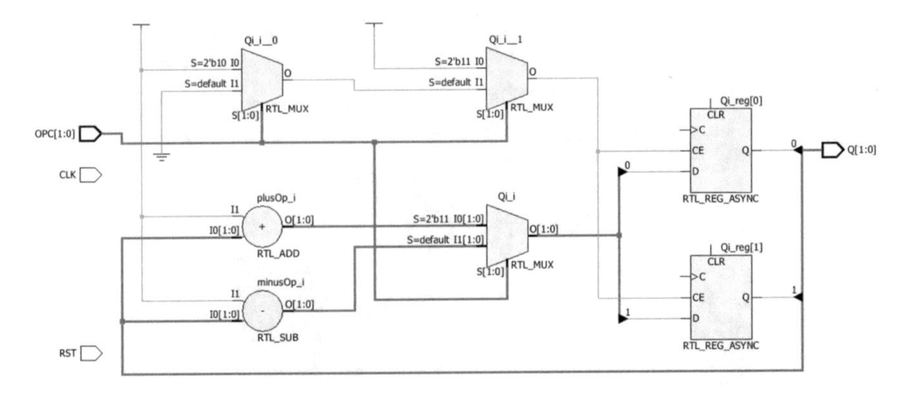


**Fig. 2.25** RTL of the ascending/descending enable counter

```
20   process(RST,CLK)
21   begin
22     if RST = '1' then
23       Qi <= (others => '0');
24     elsif rising_edge(CLK) then
25       if E = '1' then
26         Qi <= Qi(n-2 downto 0) & L;
27       else
28         Qi <= Qi;
29       end if;
30     end if;
31   end process;
32
33 end behavioral;
```

**Listing 2.21** Left-Shift register parallel output description

Other common sequential circuit is the counter. Figure 2.25 shows a RTL view of a counter ascending/descending with enable module 4. The inputs signals are: clock (CLK), asynchronous reset (RST), and operation counter (OPC). The output is the signal vector Q of 2-bit.

Ascending/descending enable counter simulation is shown in Fig. 2.26. When OPC is "00" or "01" the present value of the counter is holding. When OPC = "11" the value is increased in one each clock cycle and when OPC = "10" the value is decreased in one each clock cycle.

The description of the ascending/descending enable counter is shown in listing 2.22. The behavior of input signal OPC is described from line 24–30. In this example Qi was defined of type unsigned. Line 17 shows the output assigned, due to Qi is the type unsigned a signal conversion must be done using std_logic_vector.

```
1 library ieee;
2 use ieee.std_logic_1164.all;
3 use ieee.numeric_std.all;
4
5 entity counter is
```

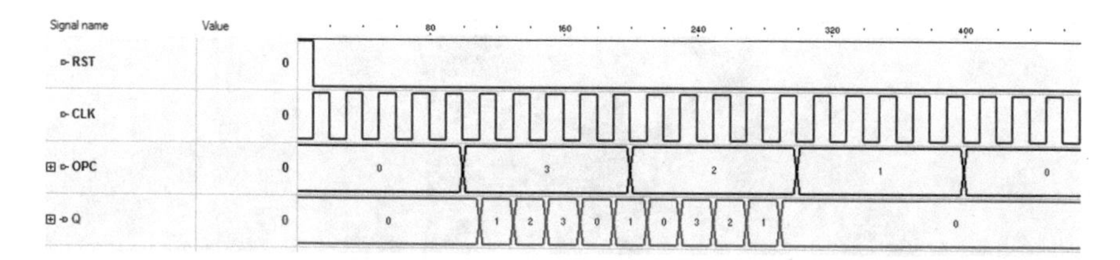


**Fig. 2.26** Simulation of the ascending/descending enable counter

```
6   port (
7   RST : in  std_logic;
8   CLK : in  std_logic;
9   OPC : in  std_logic_vector(1 downto 0);
10  Q   : out std_logic_vector(1 downto 0)
11     );
12 end counter;
13
14 architecture behavioral of counter is
15 signal Qi : unsigned(1 downto 0);
16 begin
17   Q <= std_logic_vector(Qi);
18
19   process(RST,CLK)
20   begin
21     if RST = '1' then
22       Qi <= (others => '0');
23     elsif rising_edge(CLK) then
24       if OPC = "11" then
25         Qi <= Qi + 1;
26       elsif OPC = "10" then
27         Qi <= Qi - 1;
28       else
29         Qi <= Qi;
30       end if;
31     end if;
32   end process;
33
34 end behavioral;
```

**Listing 2.22** Counter ascending/descending enable description

The last example is a finite state machine (FSM). The inputs of the FSM are: asynchronous reset (RST), clock (CLK), enable (A), and enable (B). The output is a vector of 2 bits (Y). The FSM is shown in Fig. 2.27. The FMS has four states, when the reset is active the FSM goes to state 1. For each state if signal A = ‘0’ then the FSM stays is the actual state, if A = ‘1’ and B = ‘1’ the FSM goes to the next state, for A = ”1’ and B = ‘0’ the FSM returns to the previous state. In state one (S1) the output is Y = “00”, Y = “01” for S2, Y = “10” for S3 and Y = “11” for S4. This is a Moore Machine, then, the output depends on the actual state.

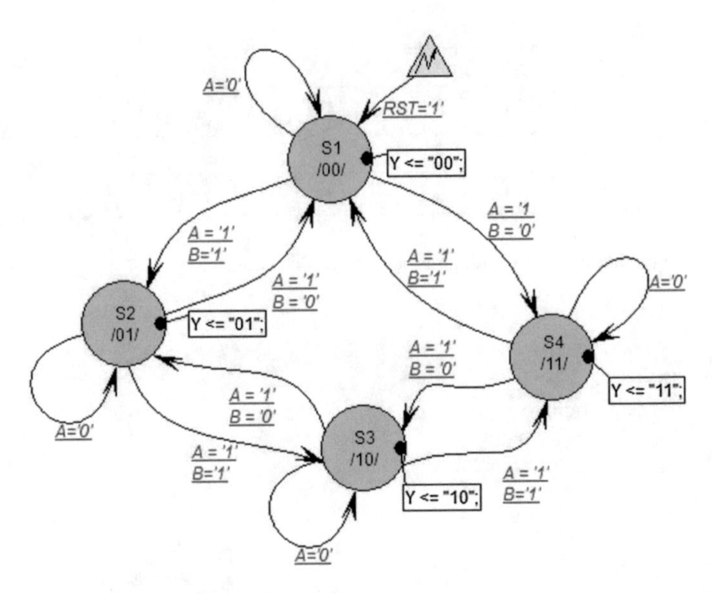


Fig. 2.27 Finite state machine

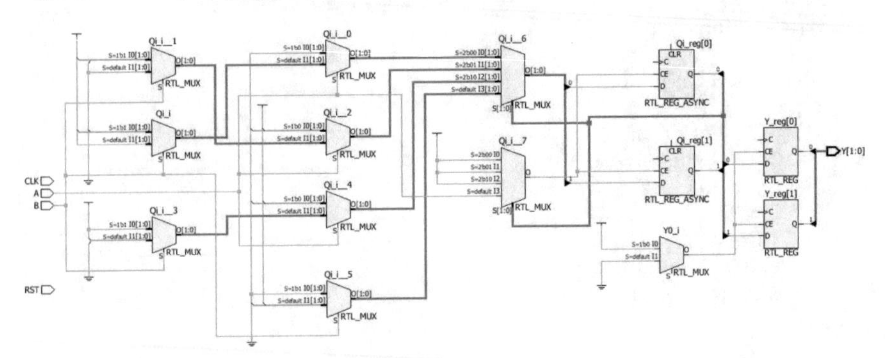


Fig. 2.28 RTL of the finite state machine

RTL view of the FSM is shown in Fig. 2.28. Simulation of the FSM is presented in Fig. 2.29, here one can see that the FSM has a behavior as the previous counter example, signal A and B represent the signal OPC in the counter.

```
1  library ieee;
2  use IEEE.std_logic_1164.all;
3
4  entity fsm is
5   port (
6   RST : in  std_logic;
7   CLK : in  std_logic;
8   A   : in  std_logic;
9   B   : in  std_logic;
10  Y   : out std_logic_vector(1 downto 0)
```

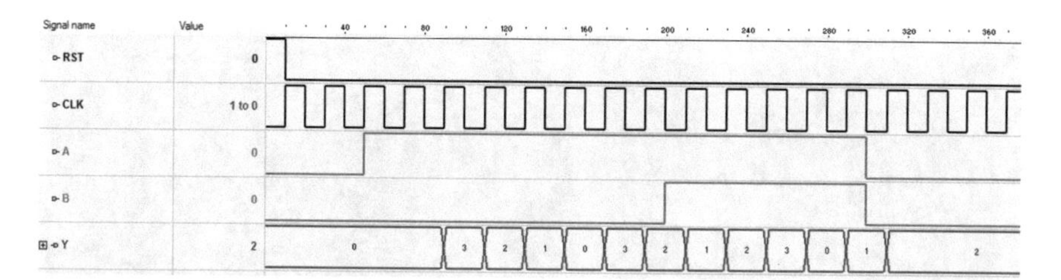


**Fig. 2.29** Simulation of the FSM

```
11      );
12 end fsm;
13
14 architecture behavioral of fsm is
15 signal Qi : std_logic_vector(1 downto 0);
16 begin
17   Y <= Qi;
18   process(RST,CLK)
19   begin
20     if RST = '1' then
21       Qi <= "00";
22     elsif rising_edge(CLK) then
23       case Qi is
24         when "00" =>
25         if A = '0' then
26           Qi <= "00";
27         elsif B = '1' then
28           Qi <= "01";
29         else
30           Qi <= "11";
31         end if;
32
33         when "01" =>
34         if A = '0' then
35           Qi <= "01";
36         elsif B = '1' then
37           Qi <= "10";
38         else
39           Qi <= "00";
40         end if;
41
42         when "10" =>
43         if A = '0' then
44           Qi <= "10";
45         elsif B = '1' then
46           Qi <= "11";
47         else
48           Qi <= "01";
49         end if;
50
51         when others =>
```

```
52            if A = '0' then
53              Qi <= "11";
54            elsif B = '1' then
55              Qi <= "00";
56            else
57              Qi <= "10";
58            end if;
59          end case;
60        end if;
61      end process;
62
63  end behavioral;
```

**Listing 2.23** Finite state machine behavioral description

The behavioral description of the FSM is presented in listing 2.23. To describe the FSM a case structure is used, for each state one case is used. The output is assigned in line 17.

# Chapter 3
# Matlab-Simulink Co-Simulation

## 3.1 Co-Simulation Active-HDL/Matlab-Simulink

Active-HDL allows the generation of block descriptions for Simulink [17]. To generate it, go to design browser window and give a right click on the file to be generated. An option menu is displayed as shown by Fig. 3.1. In this menu, click on Generate Block Description for Simulink option. Now a Simulink block of the VHDL description is obtained. In order to use it, open Matlab and then Simulink.

Figure 3.2 shows the Simulink Library Browser window, if the link between Matlab and Active-HDL was generated correctly, Active-HDL Blockset appears in this window, if not visit "www.aldec.com". to find a solution. In the Active-HDL Blockset the following blocks are contained:

- **Active-HDL Co-Sim** is a block that stores general co-simulation settings and then passes them to Active-HDL invoking its instances during the start-up of the co-simulation session.
- **HDL Black Box** is a block that represents in the Simulink environment an HDL unit that will be co-simulated using Simulink and Active-HDL.
- **HDL Black Box for DSP Builder** is a block that can be integrated with a model coming from Altera DSP Builder ver. 7.1 and higher.
- **HDL Black Box for Synplify DSP** is a block that can be integrated with a model coming from Synplify DSP ver. 3.2 and higher.
- **HDL Black Box Manager for System Generator 8.x** is a block that manages instantiation of HDL black boxes within a model coming from Xilinx System Generator ver. 8.x or newer.

Open a new Simulink Model. In this new model, add the Active-HDL Co-sim block, then give two clicks on it to configure it. Figure 3.3 shows the configuration window of the Active-HDL Co-Sim block. One important option is the Reference period in which the user selects the iteration time for Active-HDL and Simulink.



E. Tlelo-Cuautle et al., *Engineering Applications of FPGAs*,
DOI 10.1007/978-3-319-34115-6_3

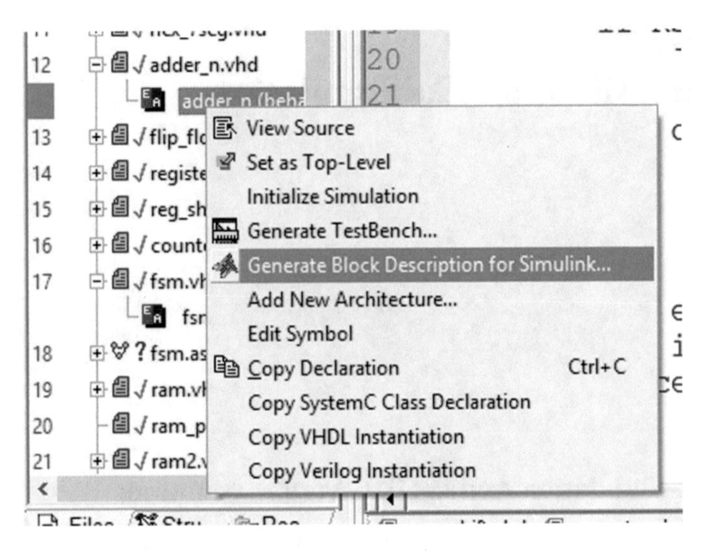


**Fig. 3.1** Generate block description for simulink

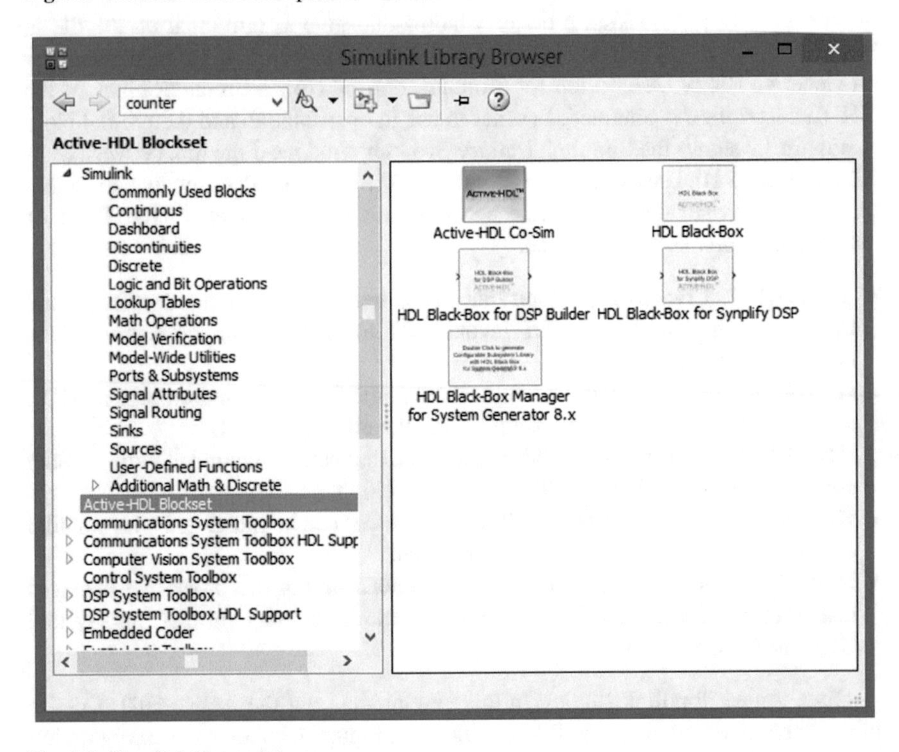


**Fig. 3.2** Simulink library browser

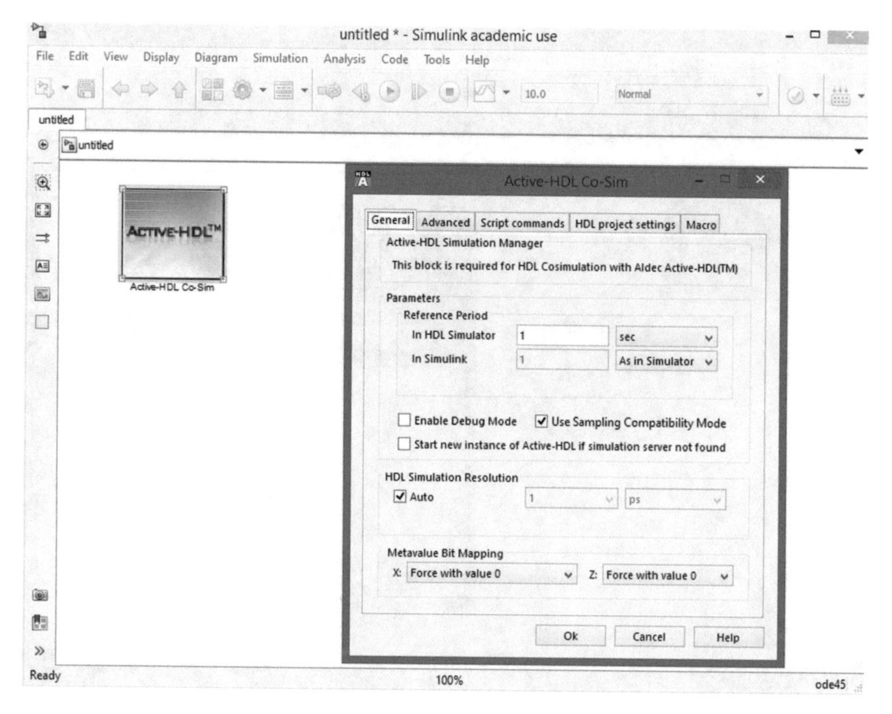


**Fig. 3.3** Active-HDL Co-Sim

Now include a HDL Black Box into the new model as shown in Fig. 3.4. A window to select one M-file is displayed. Here, the circuit to be simulated into simulink must be selected. In the image example there is just one file "adder_n", which was explained in Chap. 2. The M-files must be located into the current working directory, if not, add the directory to Matlab search path.

When the Black box is configured, the next step is the parameters selection. Figure 3.5 shows the parameters of the black box. In this case, in the option of Input Ports, for the adder_n the inputs are: A, B and, Cin. Inputs A and B are vectors of n-bit and Cin is the carry input; here the cast of each input can be changed (unsigned, signed, boolean) and some other features.

Figure 3.6 shows the output ports, for adder_n are: Sum and Cou. Like the previous case, the cast can be changed.

As can be seen in Listing 2.18, the adder has one generic signal "n" the type integer, this parameter can be modified into the parameters option as shown in Fig. 3.7. Default value in this case is four and the actual value (also four) can be modified to increase or decrease the number of bits for the adder. Then, this is an excellent option that exploit the reusability of the VHDL descriptions, so with the modification of this parameter an adder of 4 or 5 or n bits can be included into the simulation.

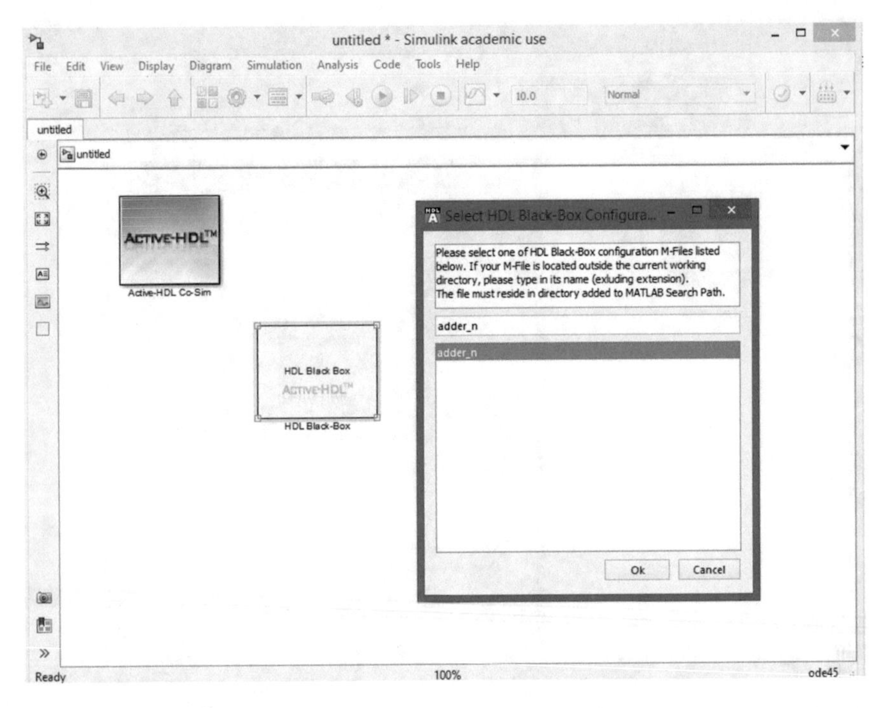


**Fig. 3.4** HDL Black Box Configuration

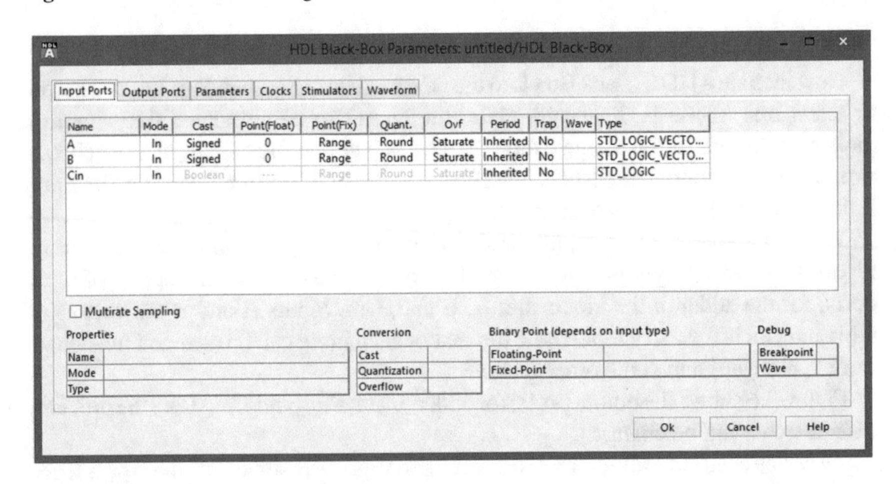


| Name | Mode | Cast | Point(Float) | Point(Fix) | Quant. | Ovf | Period | Trap | Wave | Type |
|---|---|---|---|---|---|---|---|---|---|---|
| A | In | Signed | 0 | Range | Round | Saturate | Inherited | No | | STD_LOGIC_VECTO... |
| B | In | Signed | 0 | Range | Round | Saturate | Inherited | No | | STD_LOGIC_VECTO... |
| Cin | In | Boolean | --- | Range | Round | Saturate | Inherited | No | | STD_LOGIC |

**Fig. 3.5** HDL Black Box Parameters—Input ports

If the design has clocks or any signal to be synchronous, they can be stimulated in clocks option. This option is shown in Fig. 3.8. In the case of the adder there are no clock or synchronization signals.

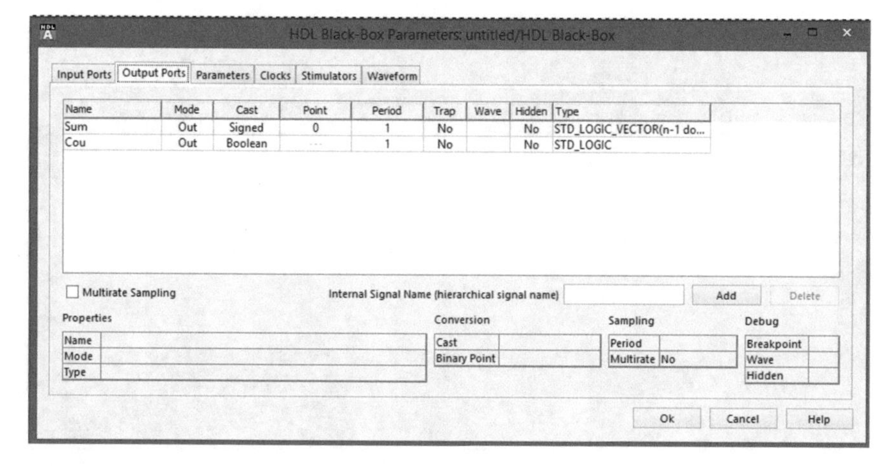


**Fig. 3.6** HDL Black Box Parameters—Output ports

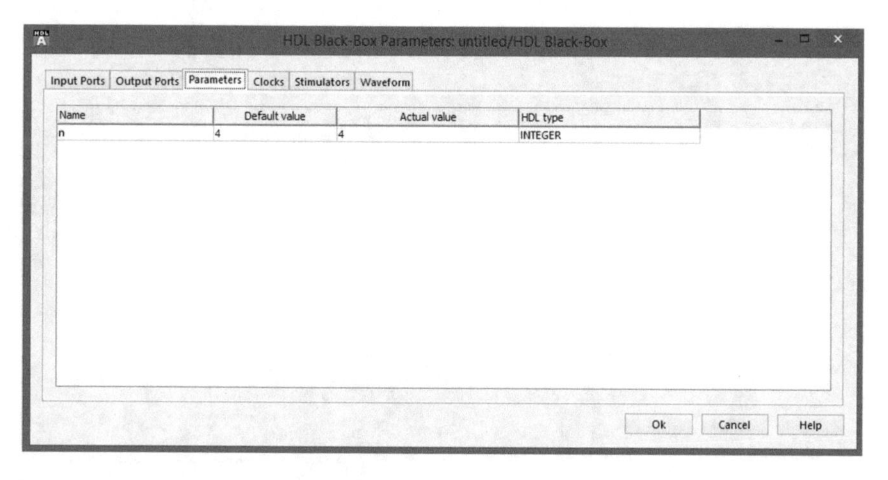


**Fig. 3.7** HDL Black Box Parameters—Parameters

Any signal can be internally stimulated, for example, an asynchronous reset can be stimulated through a formula. This stimulation is done in the stimulators option shown in Fig. 3.9. Signals can be stimulated here or using an external source (external because is not included into the black box, internal because is stimulated into the black box).

The last option corresponds to waveforms, in this option the user must select the signal to be included into the Active Simulator (Fig. 3.10). When the model is running, matlab automatically open Active and accomplish the simulation using the stimulus selected in Simulink.

Figure 3.11 shows the adder_n connected in Simulink, the inputs are stimulated with a constants and the outputs are connected to displays. Input A is 8, B = 10 and

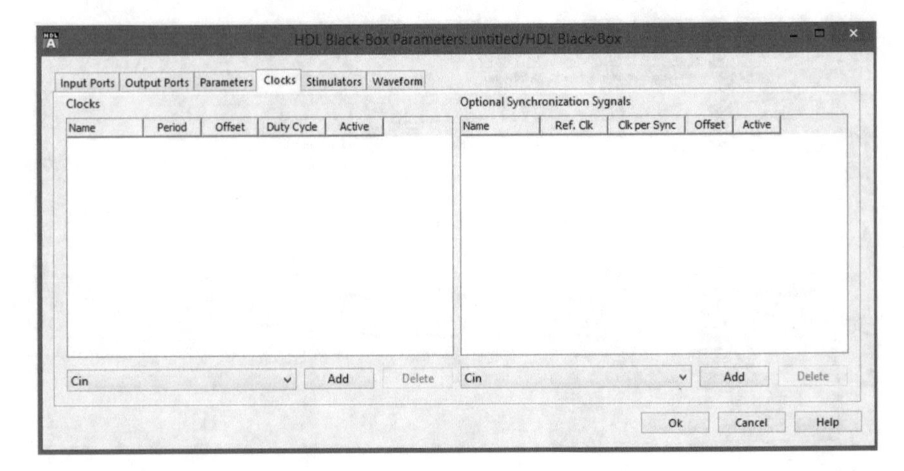


Fig. 3.8 HDL Black Box Parameters—Clocks

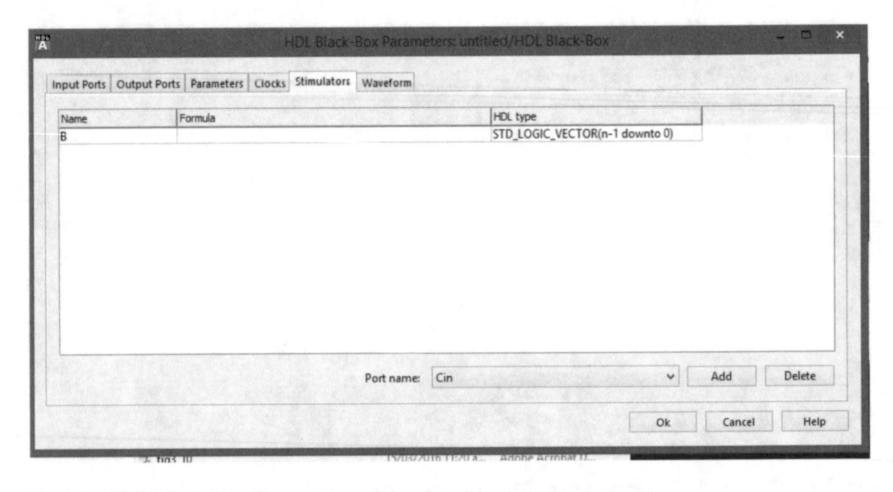


Fig. 3.9 HDL Black Box Parameters—Stimulators

Cin = 1, the outputs are: Sum = 3 and Cou = 1. In hexadecimal the output is ×13 and in decimal is 19.

Another simple example using co-simulation is presented in Fig. 3.12. In this case, the circuit to be simulated is a ROM. The ROM contains the sum of two sines. The input ADD is the ROM address, and it is stimulated using a Free-Running Counter block. The output D is connected to a Scope in order to be the signal. Figure 3.13 shows the signal displayed in the Scope and the sum of sines is clear. Figure 3.14 shows the signals input ADD and output D simulated in Active, this simulation was automatically made by Simulink. Now, any VHDL description can be included into the Simulink using the Active Block Box, once in simulink all the Matlab tools can be used to improve the simulation process.

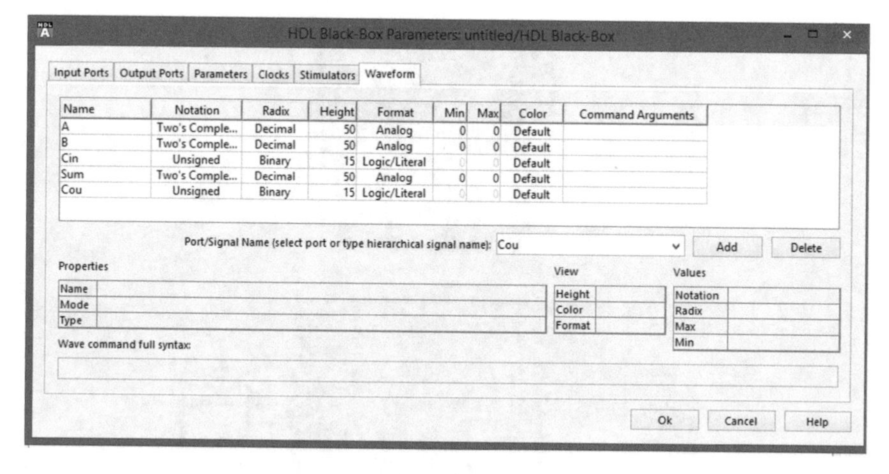


Fig. 3.10 HDL Black Box Parameters—Waveforms

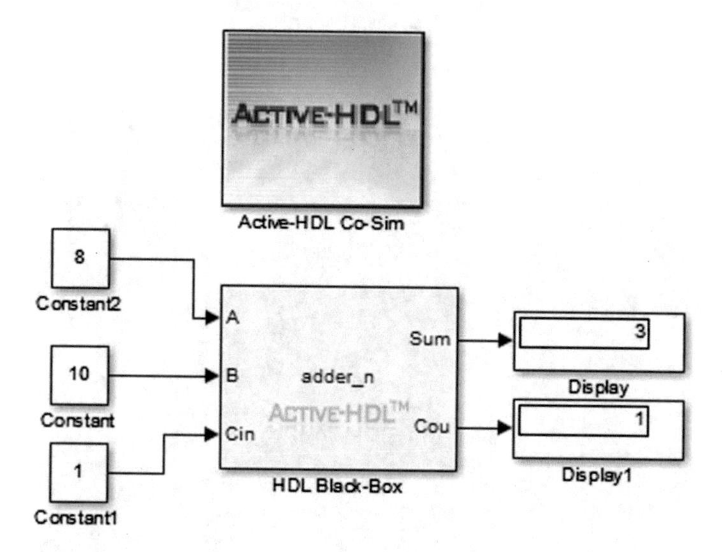


Fig. 3.11 Adder_n in Simulink

## 3.2 Co-Simulation Xilinx System Generator/Matlab-Simulink

System Generator by Xilinx allows the automatic VHDL code generation and simulation using Matlab-Simulink [18]. Similar to Active, Xilinx System Generation creates a blockset library to be used into Simulink. The blockset is shown in Fig. 3.15, its content

- **AXI4 Blocks** Includes every block that supports the AXI4 Interface.

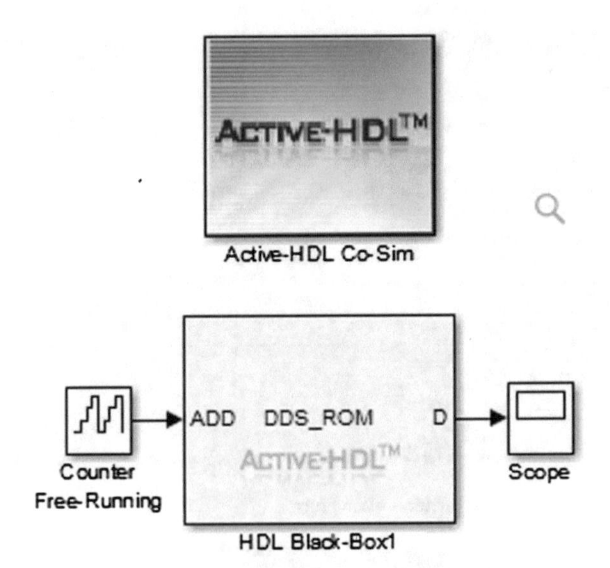


**Fig. 3.12** ROM

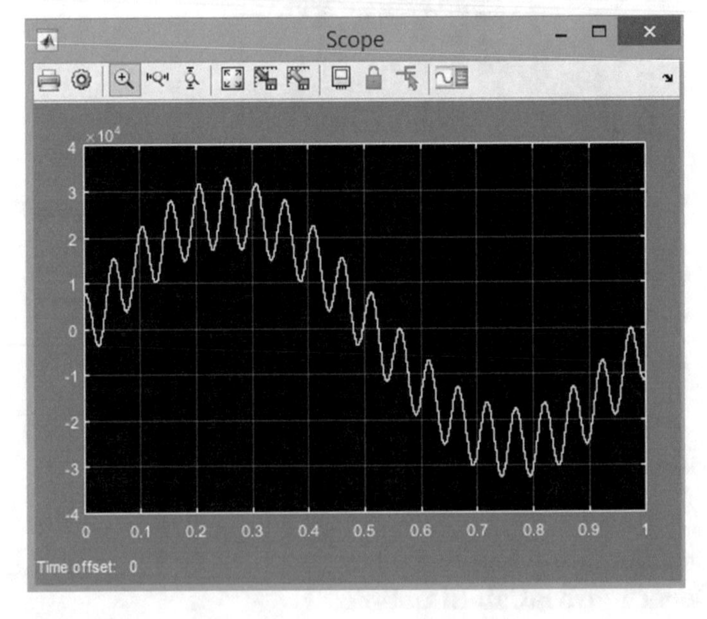


**Fig. 3.13** ROM Scope

- **Basic Elements Blocks** Includes standard building blocks for digital logic.
- **Communication Blocks** Includes forward error correction and modulator blocks, commonly used in digital communications systems.
- **Control Logic Blocks** Includes blocks for control circuitry and state machines.

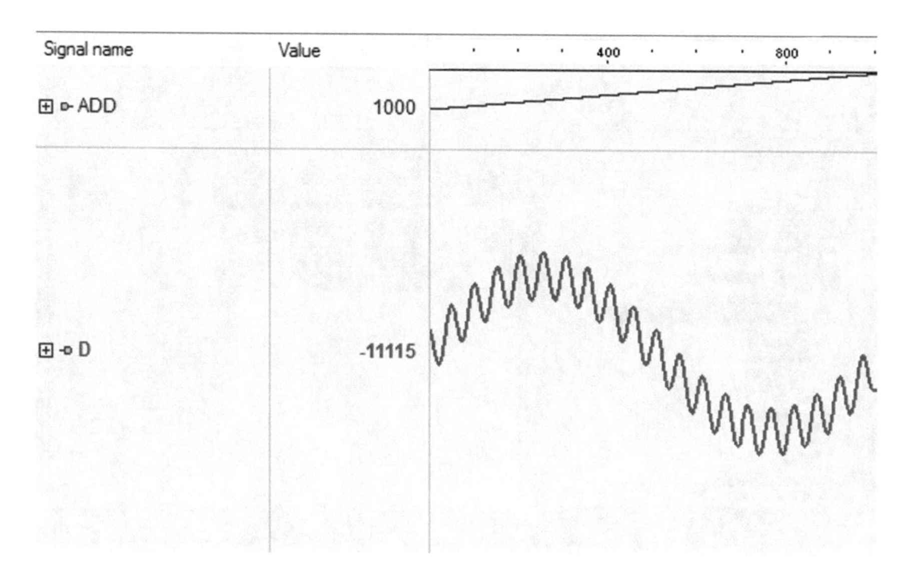


**Fig. 3.14** Active-HDL ROM Simulation

- **Data Type Blocks** Includes blocks that convert data types (includes gateways).
- **DSP Blocks** Includes Digital Signal Processing (DSP) blocks.
- **Floating-Point Blocks** Includes blocks that support the Floating-Point data types as well as other data types. Only a single data type is supported at a time. For example, a floating-point input produces a floating-point output; a fixed-point input produces a fixed-point output.
- **Index Blocks** Includes all System Generator blocks.
- **Math Blocks** Includes blocks that implement mathematical functions.
- **Memory Blocks** Includes blocks that implement and access memories.
- **Tool Blocks** Includes "Utility" blocks, e.g., code generation (System Generator token), resource estimation, HDL co-simulation, etc.

Figure 3.16 shows the blocks contained in the Xilinx DSP blockset, all the blocks with background color blue can be implemented into the FPGA and are free. The blocks with background white are for utility or tool. Green background color blocks go into the FPGA and are Licensed Cores, it is necessary to purchase the Core license in the Xilinx web.

Figure 3.17 shows the block contained in the Xilinx Math blockset, as one can see there are several mathematical functions to be used. From here some blocks will be used to show an example.

In a new simulink model, a free counter, an add, and a constant are included. These blocks were taken from the Xilinx BlockSet—Math, also two gateways out were included. An scope taken from the sinks tools is added to the model. Figure 3.18 shows the interconnection blocks. The adder is connected directly to the scope and

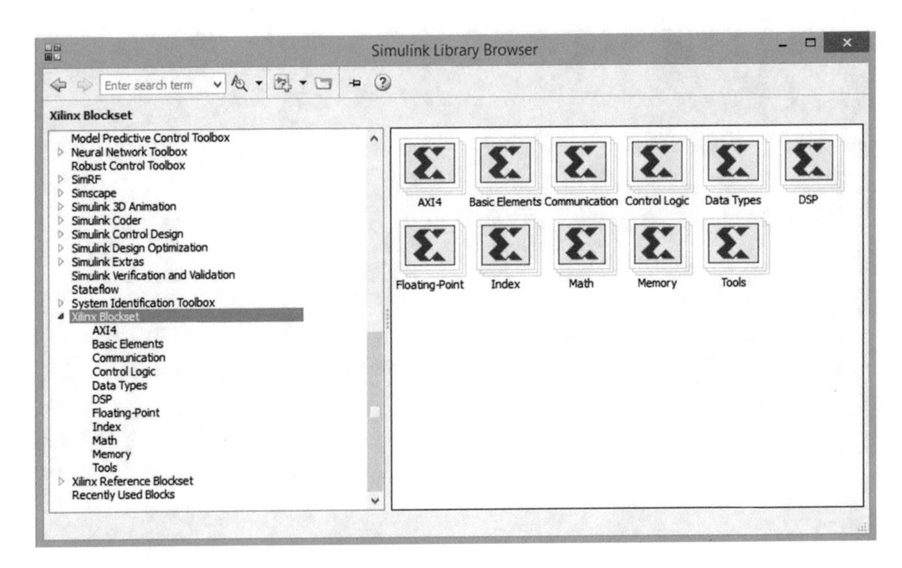


Fig. 3.15 Xilinx System Generator BlockSet

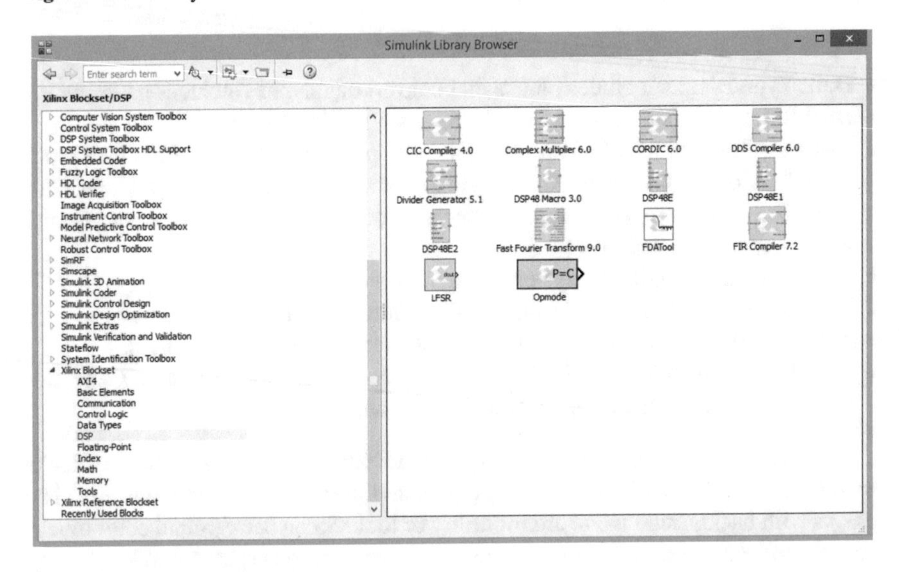


Fig. 3.16 Xilinx BlockSet—DSP

to one adder input. The second adder input is connected to a constant with a value of four. The sum is connected to the scope.

The scope is shown in Fig. 3.19, here one can see the simulation of the generated circuit. The top signal corresponds to the counter, it begins at zero and increases its value by one. The bottom signal corresponds to the adder sum, as one can see, it also

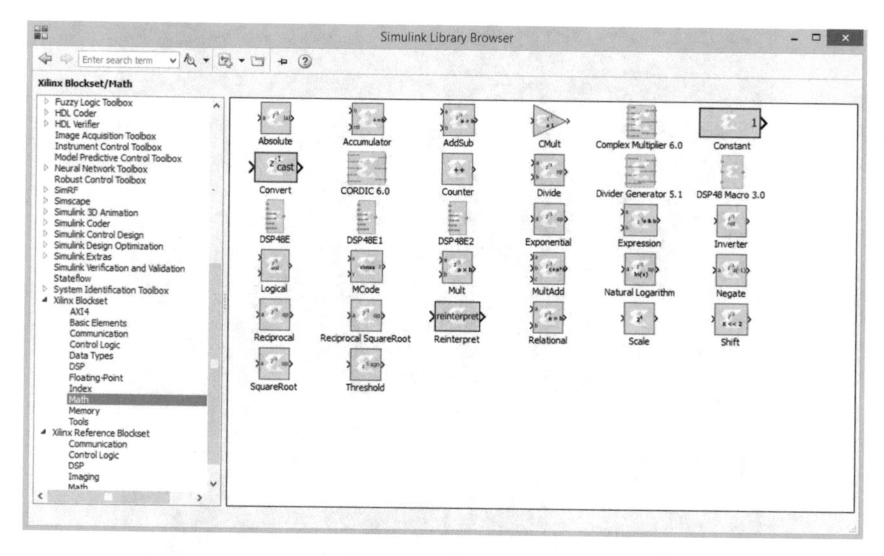


**Fig. 3.17** Xilinx BlockSet—Math

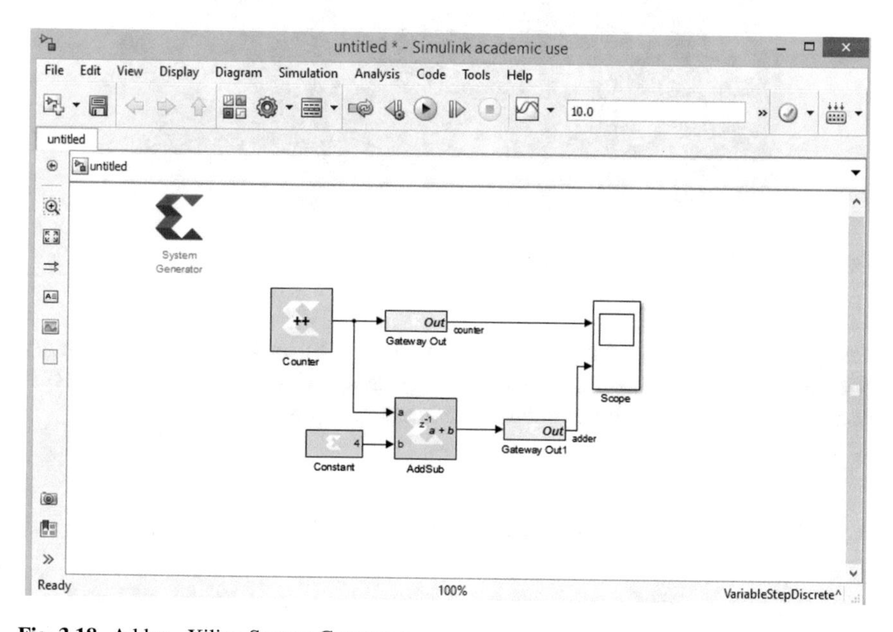


**Fig. 3.18** Adder—Xilinx System Generator

begins at zero and the sum with the constant (with a four value) is reflected in the second sample time, this is because the adder has one time delay.

In order to generate the VDHL code, double click the Xilinx System Generation icon, this allows to configure the parameters to generate the code. Figure 3.20 shows

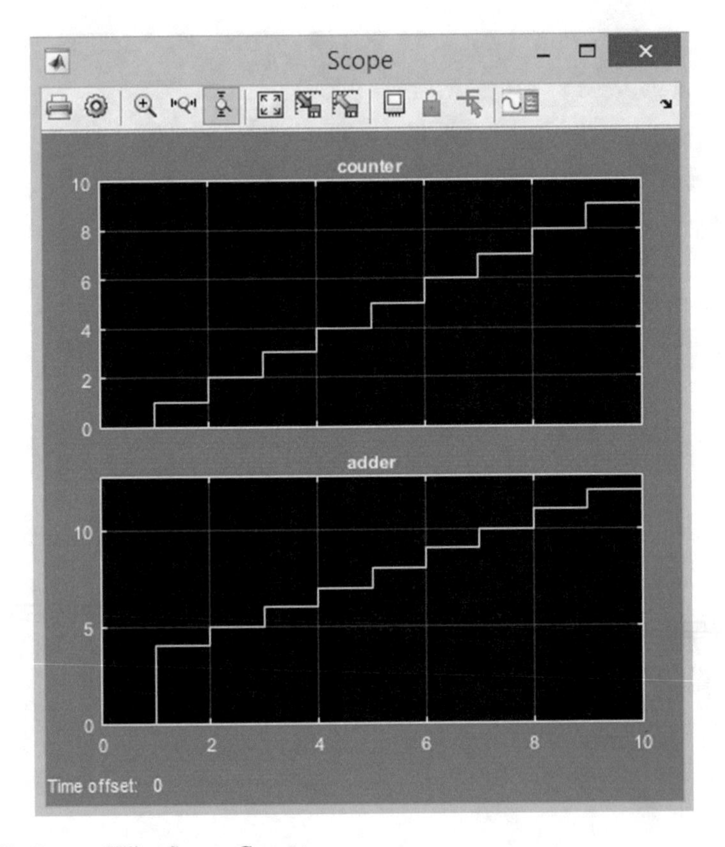


**Fig. 3.19** Scope—Xilinx System Generator

the System Generation configuration window, here the user selects the type of compilation, the FPGA to be implemented, language (VHDL or Verilog), the VHDL library, target directory, and synthesis and implementation strategy.

## 3.3 Co-Simulation Altera DSP Builder/Matlab-Simulink

Altera provides a tool for automatic code generation and Matlab-Simulink co-simulation. This tool is the Altera DSP Builder [19]. The Altera DSP Builder is shown in Fig. 3.21, which gives the following options:

- Altlab
- Arithmetic
- Boards
- Complex Type
- Gate & Control

Fig. 3.20 Xilinx System Generator

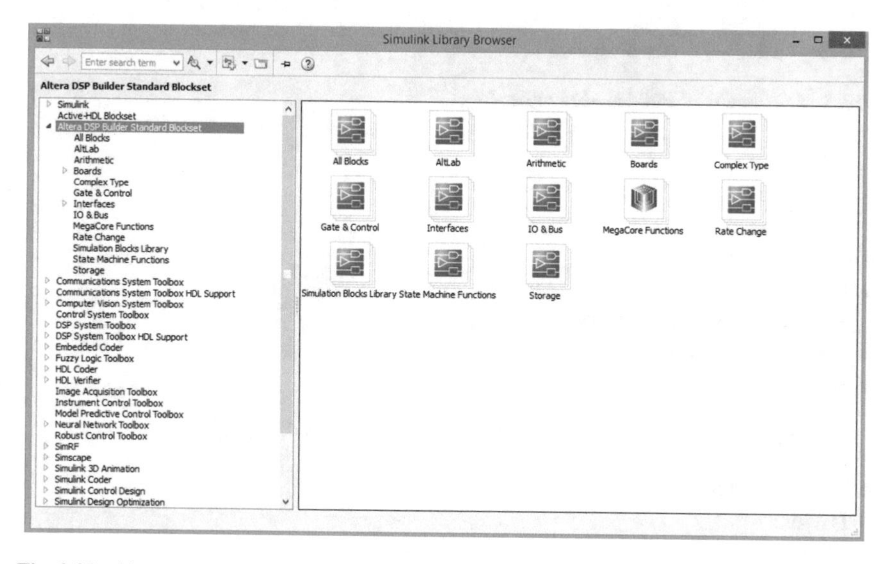


Fig. 3.21 Altera DSP Builder

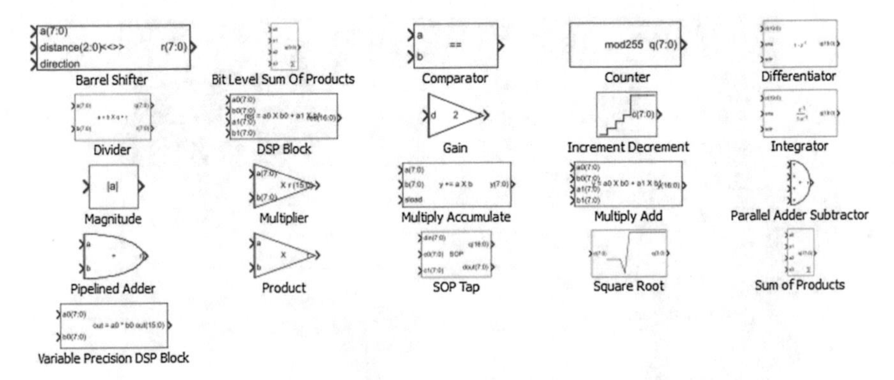


**Fig. 3.22** Altera DSP Builder—Arithmetic Blockset

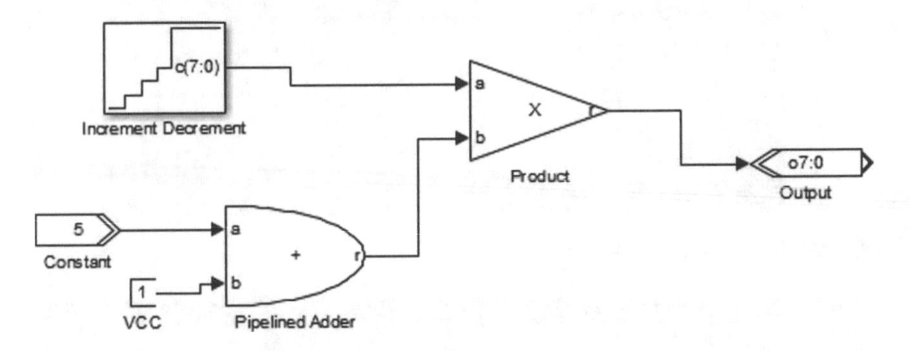


**Fig. 3.23** Altera DSP Builder—Example

- Interfaces
- IO & Bus
- MegaCore Functions
- Rate Change
- Simulation Blocks Library
- State Machine Functions
- Storage

Figure 3.22 shows the arithmetic blockset contents, which include

- Barrel Shifter
- Divider
- Magnitude
- Pipelined Adder
- Variable precision DSP block
- Product
- Multiplier
- DSP Block
- Bit level Sum of Products

- Comparator
- Gain
- Multiply accumulate
- SOP Tap
- Square Root
- Multiply Add
- Increment Decrement
- Counter
- Differentiator
- Integrator
- Parallel adder subtractor
- Sum of products

As one can be realized, the Altera DSP Builder has a lot of tools to develop a complete design. A simple example is shown in Fig. 3.23.

Matlab-Simulink simulation of the Altera DSP Builder simulation and its implementation using an Altera Board are left to the reader.

# Chapter 4
# Chaos Generators

## 4.1 On Piecewise-Linear (PWL) Functions

This section shows the description of three mathematical models associated to piecewise-linear (PWL) functions that can be used in a dynamical system to generate from double-scroll to multi-scroll chaotic attractors. Those PWL functions can be applied to a third-order dynamical system, can be increased in a systematic way, and can be symmetric or nonsymmetric.

### *4.1.1 Saturated Function Series as PWL Function*

A continuous-time chaotic oscillator can be modeled by (4.1), where $x$, $y$ and $z$ are the state variables with coefficients $a$, $b$, $c$ and $d_1$, which are constants and take values in the interval [0, 1], and also includes a PWL function $f(x; q)$ that can be implemented by saturated function series. For instance, $f(x; q)$ can be increased in number of saturated levels, which are associated to the number of scrolls being generated. In discrete electronics, $f(x; q)$ can be implemented using comparators [21], but they have low-frequency response [22–24], and it can be implemented much better using digital circuitry, e.g., FPGAs [25]. In this dynamical system, the PWL function depends on the state variable $x$, and it can be described by linear segments as shown in (4.2), where two saturated levels are generated, as shown in Fig. 4.1, and where $q_1$ denotes the break-points and $k_1$ the saturation levels, so that the slope beings $2k_1/2q_1 = k_1/q_1$.

$$\begin{aligned}\dot{x} &= y\\ \dot{y} &= z\\ \dot{z} &= -ax - by - cz + d_1 f(x; q)\end{aligned} \tag{4.1}$$



E. Tlelo-Cuautle et al., *Engineering Applications of FPGAs*,
DOI 10.1007/978-3-319-34115-6_4

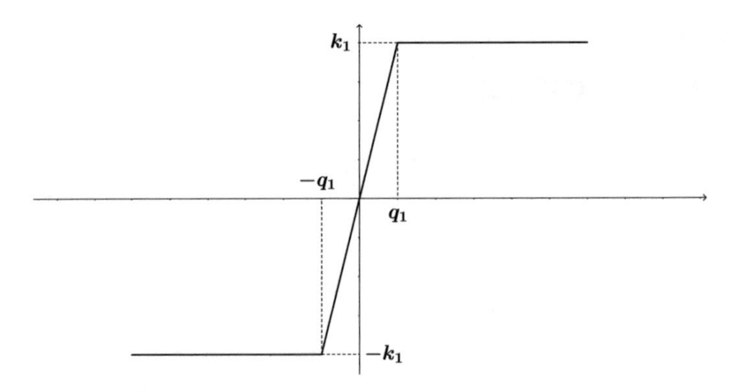


**Fig. 4.1** 3-segments PWL function based on saturated function series

$$f_0(x;q) = \begin{cases} k_1, & \text{if } x > q_1 \\ \frac{k_1}{q_1}x, & \text{if } |x| \le q_1 \\ -k_1, & \text{if } x < -q_1, \end{cases} \tag{4.2}$$

This PWL function can be extended and shifted to generate odd and even numbers of saturation levels that are related to the number of scrolls to be generated. Thinking on design automation, one can describe the PWL function by introducing break-points and saturation levels from the left to the right side on the horizontal axes. The slopes can be evaluated using these values, so that one can describe symmetrical and nonsymmetrical PWL functions.

### *4.1.2 Chua's Diode as PWL Function*

Lets us consider the dynamical system described by three state variables, as given by (4.3), where $f(x)$ is a PWL function consisting of negative slopes also known as Chua's diode [26]. Depending on the number of scrolls being generated, the negative slopes increase in a systematic way for odd and even number of scrolls. Figure 4.2 shows a PWL function with negative slopes $m_1$ and $m_2$ and break-points $\pm b_1$. That PWL function can generate 2-scrolls and more even number of scrolls can be generated by increasing the break-points and retaining the same slopes $m_1$ and $m_2$.

In (4.3), $x$, $y$, and $z$ are the state variables, $\alpha$ and $\beta$ are coefficients that are real numbers in general, and $f(x)$ models Chua's diode as a PWL function with negative slopes that can be described by (4.4). It is function of the state variable $x$ and of the real constants $m_i$ and $b_i$ that define the slopes and break-points to generate $n$ number of scrolls.

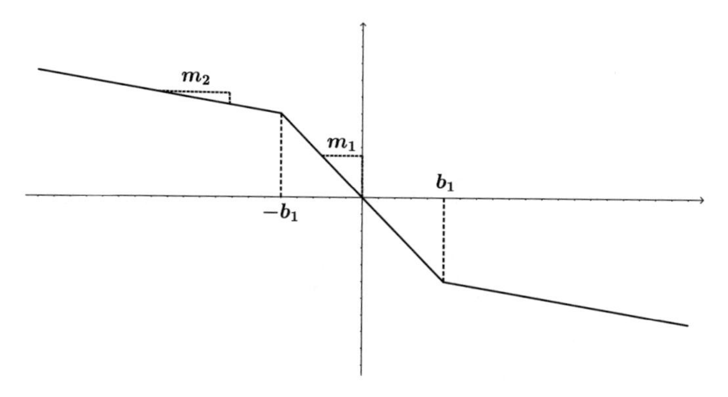


**Fig. 4.2** 3-segments PWL function based on negative slopes, also known as Chua's diode

$$\begin{aligned}\dot{x} &= \alpha(y - x - f(x))\\ \dot{y} &= x - y + z\\ \dot{z} &= -\beta y\end{aligned} \tag{4.3}$$

$$f(x) = m_{2n-1}x + \frac{1}{2}\sum_{i-q}^{2n+1}(m_{i-1} - m_i)(|x + b_i| - |x - b_i|) \tag{4.4}$$

### *4.1.3 Sawtooth as PWL Function*

Chua's circuit can also be modeled by three state variables and a sawtooth as the PWL function [27], as given by (4.5) and shown in Fig. 4.3. $x$, $y$ and $z$ are the state variables, and again, as for the dynamical system based on Chua's diode, $\alpha$ and $\beta$ are coefficients that are real numbers in general. In this case, $f(x)$ models a PWL function with sawtooth behavior consisting of positive slopes described by (4.6) and (4.7) for generating even and odd number of scrolls, respectively.

$$\begin{aligned}\dot{x} &= \alpha(x - f(x))\\ \dot{y} &= x - y + z\\ \dot{z} &= -\beta y\end{aligned} \tag{4.5}$$

$$f_1(x) = \xi\{x - A_1[-\text{sgn}(x) + \sum_{i=0}^{N-1}(sgn(x + 2iA_1) + sgn(x - 2iA_1))]\} \tag{4.6}$$

$$f_1(x) = \xi\{x - A_1[-\text{sgn}(x) + \sum_{i=0}^{N-1}(sgn(x + (2i + 1)A_1) + sgn(x - (2i + 1)A_1))]\} \tag{4.7}$$

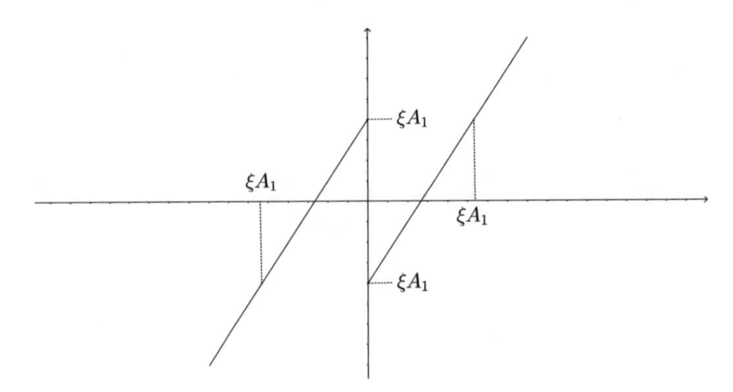


**Fig. 4.3** 3-segments PWL function based on a sawtooth function

## 4.2 On the Simulation of Chaos Generators for FPGA Implementation

Chaotic systems modeled by continuous or discrete equations are solved or simulated using digital hardware resources, e.g., the majority of them using computers. That way, simulating dynamical systems for large times may lead to non-convergence to the desired solution. However, as the number of bits that are processed by a computer or a digital hardware is finite, such a problem is a big challenge and one just knows that some errors are present. Detailed comments are given in [28], where one can appreciate the following sentence: due to inherent properties of digital computers results found by numerical simulations are almost never exact. Nevertheless, computer-generated solutions are often accepted as true solutions.

In this manner, from what is demonstrated in [28], one can implement continuous chaotic oscillators using FPGAs, just to verify or observe chaotic behavior. In addition, guaranteeing such chaotic behavior in a very large time is a hard open challenge.

Another challenge is the selection of the numerical method to solve a dynamical system of equations. One must keep in mind that the use of simple integration methods leads to reduced hardware resources in FPGAs, but the rounding errors are present more than by applying more robust integration methods that can have variable order and/or variable step size. However, if one guarantees that the solution converges, it does not matters what kind of method is used, the chaotic behavior will be similar, but the problem on rounding errors will always be present. Below, we show that the solution of a dynamical system like the ones listed above converges when using Forward Euler method. In this manner, the simulated dynamical system can be implemented in an FPGA.

The one-step method known as Forward Euler, can be applied for solving a dynamical system described in the general form by (4.8). In its discrete form it is given by (4.9), where $\phi(t, y, h)$ is known as an increment function and $\lambda = \left(\frac{\partial f}{\partial y}\right)_n$ [29, 30].

$$\frac{dy}{dt} = f(t, y) = \lambda y, \qquad y(t_0) = y_0, \qquad t \in [t_0, b] \tag{4.8}$$

$$y_{n+1} = y_n + h\phi(t_n, y_n, h), \qquad n = 0, 1, ..., N-1 \tag{4.9}$$

Assuming that (4.8) has a unique solution in $[t_0, b]$ and that $y(t) \in C^{(p+1)}\,[t_0, b]$ for $p \geq 1$, then the solution $y(t)$ can be expanded (Taylor's series) at any point $t_n$ like in (4.10). Considering (4.11) and that $h\phi(t_n, y_n, h)$ is obtained from $h\phi(t_n, y(t_n), h)$, then the approximated value for $y_n$ instead of the exact value of $y(t_n)$ is obtained from *Taylor's series method* of order $p$ and described by (4.12) to approximate $y(t_{n+1})$ [30].

$$y(t_{n+1}) = y(t_n) + hy'(t_n) + \frac{h^2}{2!}y''(t_n) + ... \tag{4.10}$$

$$h\phi(t_n, y(t_n), h) = hy'(t_n) + \frac{h^2}{2!}y''(t_n) + ... + \frac{h^p}{p!}y^{(p)}(t_n) \tag{4.11}$$

$$y_{n+1} = y_n + h\phi(t_n, y_n, h), \qquad n = 0, 1, 2, ..., N-1 \tag{4.12}$$

Therefore, if $p = 1$ one gets *Euler's method*, as described by (4.13)

$$y_{n+1} = y_n + hf(t_n, y_n), \qquad n = 0, 1, 2, ..., N-1 \tag{4.13}$$

From an stability analysis considering (4.8) and assuming that $\partial f/\partial y$ is relatively invariant in the region of interest, then the solution to (4.8) is given by (4.14), and considering that $t = t_0 + nh$ one gets (4.15),

$$y(t) = y(t_0)e^{\lambda(t-t_0)} \tag{4.14}$$

$$y(t_n) = y(t_0)e^{\lambda nh} = y_0(e^{\lambda h})^n \tag{4.15}$$

When applying a one-step method to (4.8), the solution is given by (4.16), where $c_1$ is a constant and $E(\lambda h) \approx e^{\lambda h}$ [30].

$$y_n = c_1(E(\lambda h))^n \tag{4.16}$$

Considering the set of equations for the chaotic oscillator shown in (4.17), it has a unique equilibrium point at (0, 0, 0), then its characteristic equation is given by (4.18) [31],

$$\begin{aligned}\dot{x} &= y\\ \dot{y} &= z\\ \dot{z} &= -ax - by - cz + d_1 f(x_1; h_1, p_1, q_1)\end{aligned} \tag{4.17}$$

$$\lambda^3 + c\lambda^2 + b\lambda + a = 0 \tag{4.18}$$

Solving (4.18) one gets: $\lambda_1 = -\eta$ and $\lambda_{2,3} = \alpha_1 \pm \beta i$, where $\lambda_1 < 0, \alpha_1 > 0$ and $\beta \neq 0$. Then, (4.17) has a negative eigenvalue and one pair of complex eigenvalues with positive real part [31]. As a result, a numerical method solving (4.17) will simulate chaotic behavior under the conditions of $\lambda_1, \alpha_1, \beta$. This result is extended to a multi-scroll chaotic oscillator because in (4.17), $f(x_1; , h_1, p_1, q_1)$ is a PWL function consisting of slopes, offsets and saturation values. For example, simulating the generation of 2-scrolls, the PWL function is given by (4.19), where $\alpha = 0.0165$, $m = 60.606$, $a = b = c = d_1 = 0.7$. In this case: $\lambda_1 = -0.8410142$, $\lambda_{2,3} = 0.07050 \pm 0.90201i$.

$$f(x) = \begin{cases} 1 & \text{if } x > \alpha \\ \frac{x}{m} & \text{if } |x| \leq \alpha \\ -1 & \text{if } x < -\alpha \end{cases} \tag{4.19}$$

Since (4.17) has at least one $\lambda > 0$, then Euler's method is relatively stable if $| E(\lambda h) | \leq e^{\lambda h}$ for $\lambda > 0$. Therefore, applying (4.13) for solving (4.17) will lead to the true solution if for an arbitrary initial condition $y_0$,

$$\lim_{h \to 0} y_n = y(t) \text{ for } t \in [t_0, b], \qquad t_n = t \tag{4.20}$$

Finally, considering the sampling theorem, $\tau_{min} = \frac{1}{f}$, the condition is $h \leq \frac{\tau_{min}}{2}$ in order to the numerical method to converge to the true solution when $h \to 0$, i.e., better when $h$ is relatively low, as it is in all the cases simulated in this book.

### 4.2.1 *One-Step Methods for Simulating the Generation of 2-Scrolls*

Forward Euler is the simple predictive method for solving initial value problems. The iterative equation is given by (4.21), where $\phi$ is the function of $y$, $y_n$ is the initial value of the state variable, $y_{n+1}$ is the computed value, and $h$ is the step size that must be small as demonstrated above.

$$y_{n+1} = y_n + \phi h \tag{4.21}$$

Another one-step and more accurate method is Runge–Kutta of higher order. For instance, the fourth order (Runge–Kutta) is given by (4.22), in which $k_1$, $k_2$, $k_3$ and $k_4$ are evaluated by (4.23).

$$y_{n+1} = y_n + \frac{1}{6}h(k_1 + k_2 + k_3 + k_4) \tag{4.22}$$

$$\begin{aligned} k_1 &= f(x_n, y_n) \\ k_2 &= f(x_n, \frac{1}{2}h, y_n + \frac{1}{2}hk_1) \\ k_3 &= f(x_n, \frac{1}{2}h, y_n + \frac{1}{2}hk_2) \\ k_4 &= f(x_n, \frac{1}{2}h, y_n + \frac{1}{2}hk_3) \end{aligned} \tag{4.23}$$

As one can infer from (4.22), the hardware realization requires multipliers, dividers, adders and subtractors. At the simulation level this is not an issue, because the computer just computes, but thinking on hardware implementations, e.g., on using FPGA, the available resources are limited and this is an issue for minimizing hardware resources or number of arithmetic operations.

By applying the Forward Euler method, (4.24)–(4.26) describe the discretization of the dynamical systems given in (4.1), (4.3) and (4.5), respectively. Those dynamical systems embed PWL functions to generate 2-scrolls.

$$\begin{aligned} x_{n+1} &= x_n + hy_n \\ y_{n+1} &= y_n + hz_n \\ z_{n+1} &= z_n + h(-ax - by - cz + d_1 f(x_n; q)) \end{aligned} \tag{4.24}$$

$$\begin{aligned} x_{n+1} &= x_n + h\alpha(y_n - x_n - m_1 x_n - \frac{1}{2}(m_0 - m_1)(|x_n + b| - |x_n - b|)) \\ y_{n+1} &= y_n + h(x_n - y_n + z_n) \\ z_{n+1} &= z_n + h(-\beta * y) \end{aligned} \tag{4.25}$$

$$\begin{aligned} x_{n+1} &= x_n + h(\alpha(x_n - \xi(x_n - A_1 \mathrm{sgn}(x_n)))) \\ y_{n+1} &= y_n + h(x_n - y_n + z_n) \\ z_{n+1} &= z_n + h(-\beta y) \end{aligned} \tag{4.26}$$

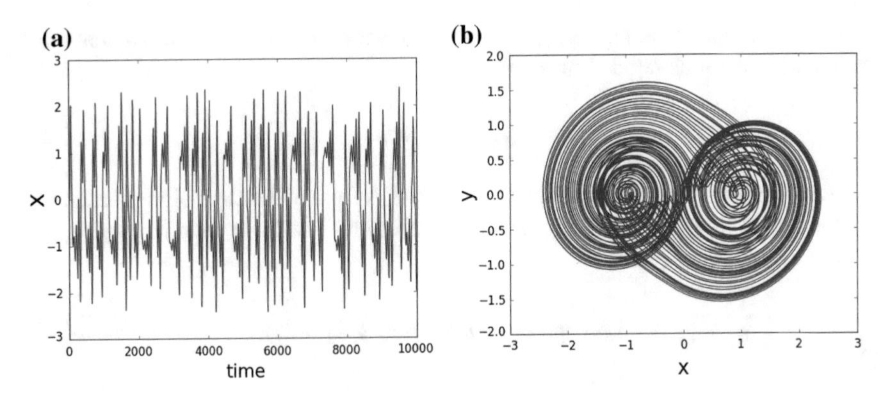


**Fig. 4.4** Chaos generator using saturated function series as PWL function. **a** State variable $x$. **b** Portrait $x$–$y$

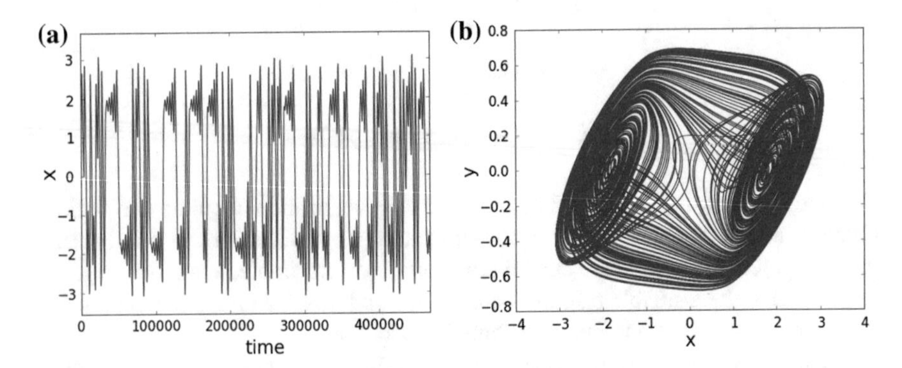


**Fig. 4.5** Chaos generator using Chua's diode as PWL function. **a** State variable $x$. **b** Portrait $x$–$y$

Figure 4.4 shows the simulation of the chaotic oscillator based on saturated function series. The state variable $x$ is shown on the left and the phase-space portrait between state variables $x$–$y$ on the right. Figure 4.5 shows the behavior of the state variable $x$ for the Chua's chaotic oscillator based on Chua's diode. Finally, Fig. 4.6 shows Chua's chaotic oscillator using sawtooth as PWL function. In all cases the phase-space portraits are shown between the state variables $x$–$y$.

## 4.3 Symmetric and Nonsymmetric PWL Functions

Chaotic oscillators based on PWL functions are relatively easier to implement than those based on continuous and nonlinear functions. Those PWL functions can be symmetric and nonsymmetric. For instance, for symmetric PWL functions based on saturated functions series one can increase the number of linear segments in a systematic way, because the slopes are the same and the saturated levels are shifted

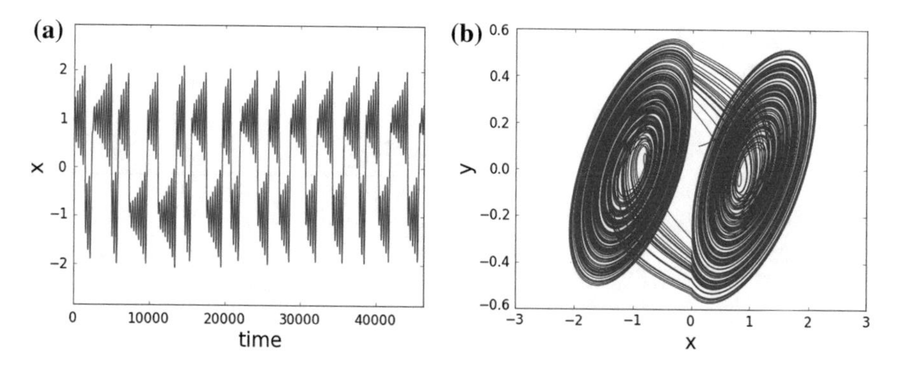


**Fig. 4.6** Chaos generator using sawtooth as PWL function. **a** State variable $x$. **b** Portrait $x$–$y$

with a multiplicative constant value. It is similar for Chua's diode, that can have 2 or 3 slopes depending if one is generating an even or odd number of scrolls, respectively. The sawtooth increases with the same slope and same shift of them. Those PWL functions can be modified to compute a high value of the MLE, as well as by varying the coefficient values [32].

If one is interested in changing the symmetry of the PWL functions, the slopes, saturation levels and shift values should be different among them, and it may increase the MLE value.

## *4.3.1 Symmetric PWL Function*

Considering the chaotic oscillator described by (4.1), one can vary the coefficients $a$, $b$, $c$ and $d_1$, to search for a high MLE value. One can also vary the break-points $q$ and shift all saturated levels. Figure 4.7 shows 5 segments with 3 saturated levels to generate 3-scrolls, which can be described by (4.27). From this PWL function, one can add more segments in both quadrants (1 and 3) to generate odd number of scrolls. To generate even number of scrolls, 2 saturated levels are required, which can be increased to generate more than 2 but even scrolls. See for example Fig. 4.8 that has 7 segments with 4 saturated levels to generate 4-scrolls, just by increasing the PWL function from Fig. 4.1. Those PWL functions are mathematically described in a similar way from (4.27).

$$f_0(x;q) = \begin{cases} k_1, & \text{if } x > q_2 \\ \frac{k_1-k_0}{q_2-q_1}x, & \text{if } q_1 \le x \le q_2 \\ k_0, & \text{if } |x| < q_1 \\ \frac{k_0-k_1}{-q_1+q_2}x, & \text{if } -q_1 \ge x \ge -q_2 \\ -k_1, & \text{if } x < -q_2, \end{cases} \tag{4.27}$$

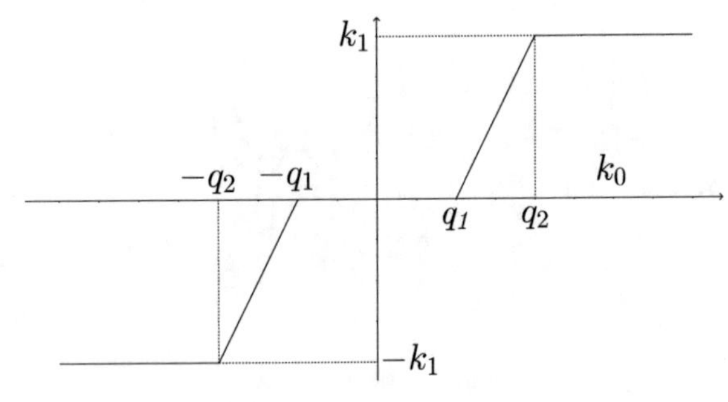


**Fig. 4.7** PWL function to generate a 3-scrolls chaotic attractor

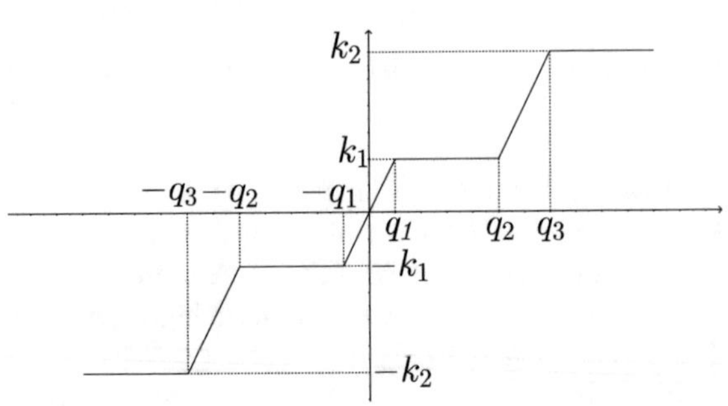


**Fig. 4.8** PWL function to generate a 4-scrolls chaotic attractor

The PWL functions given above can be increased in a systematic way to generate symmetric functions. In such a case, the values for $q_i$ and $k_i$ can be specified with the same shift values so that all the slopes are always equal to $k/q$. As examples: to generate 2-scrolls, the PWL function can have $|q_1| = 0.0165$, and $|k_1| = 1$, leading to a slope equal to 0.0165. Equation (4.28) shows an easy way to set saturation levels $k_i$ for generating even number of scrolls, and (4.29) lists the break-points $q_i$, which are also symmetrical and increased in a systematic way. For generating odd number of scrolls, the saturated levels can be set as shown by (4.30), and the corresponding break-points can be set as shown by (4.31).

$$k_i = \{..., -7, -5, -3, -1, 1, 3, 5, 7, ...\} \tag{4.28}$$

$$q_i = \{..., -2.0165, -1.9835, -0.0165, 0.0165, 1.9835, 2.0165, ...\} \tag{4.29}$$

$$k_i = \{..., -8, -6, -4, -2, 0, 2, 4, 6, 8, ...\} \tag{4.30}$$

$$q_i = \{..., -2.9835, -1.0165, -0.9835, 0.9835, 0.0165, 1.9835, ...\} \tag{4.31}$$

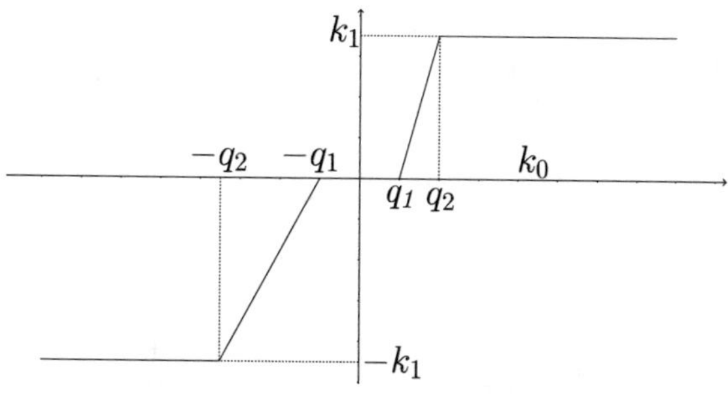


**Fig. 4.9** Nonsymmetric PWL function having 5 segments as in Fig. 4.7, but with different slopes and duration of the saturated levels

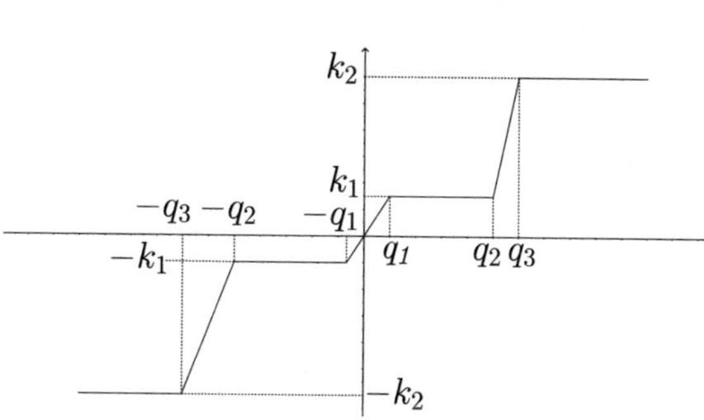


**Fig. 4.10** Nonsymmetric PWL function having 7 segments as in Fig. 4.8, but with different slopes and duration of the saturated levels

### *4.3.2 Nonsymmetric PWL Function*

When there is not a possibility of mathematically representing both the saturated levels $k_i$ and the break-points $q_i$, the PWL function is nonsymmetric. This means that both the slopes and the duration of the saturated levels are not the same. That way, the PWL functions in Figs. 4.7 and 4.8 can be nonsymmetric ones by varying the values of either or both $k_i$ and $q_i$.

One example is shown in Fig. 4.9 for generating 3-scrolls. Another example is shown in Fig. 4.10 for generating 4-scrolls. When plotting the phase-space portrait, the scrolls have not the same shape. However, to observe the desired number of scrolls being generated, one should verify that the state variables have trajectories crossing all break-points. Some examples are given in the following sections.

### *4.3.3 VHDL Simulation and Computer Arithmetic Issues*

Before creating the VHDL code for simulating a chaotic oscillator, one must establish the correct computer arithmetic, as described in Chap. 1. In this manner, by adopting fixed-point notation, the format of the digital word depends on the ranges of the values for the coefficients, PWL functions and amplitude of the state variables. It could be possible to divide or scale high values to have ranges between $\pm 1$, $\pm 10$, $\pm 20$, and so on. Some examples are listed in Table 4.1, where 28 bits can be represented in a

**Table 4.1** Fixed-point formats for a 28-bits word

| Format | Sign | Integer part | Fractional part |
|---|---|---|---|
| 4.24 | 1 bit | 3 bits | 24 bits |
| | 0 | 000 | .000000000000000000000000 |
| 5.23 | 1 bit | 4 bits | 23 bits |
| | 0 | 0000 | .00000000000000000000000 |

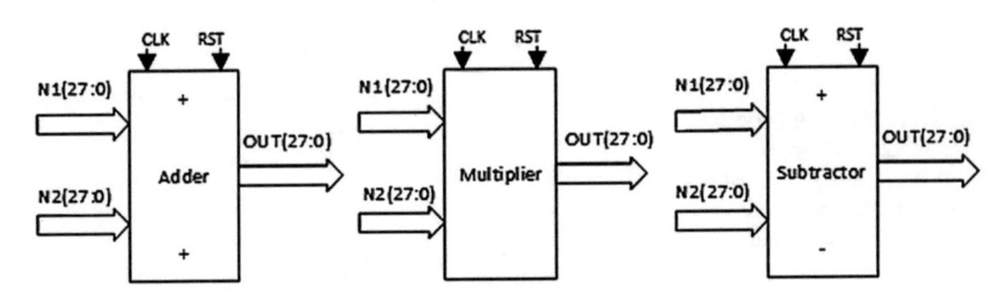


**Fig. 4.11** Block description of the adder, multiplier and subtractor processing 28 bits, and showing pins for reset and clock

4.24 format or 5.23 if the amplitude varies a little bit higher. In any case, one must consider or estimate the high value that a chaotic oscillator can generate from the operations like multiplication, addition, subtraction, etc.

Once the computer arithmetic is established, the hardware implementation is performed according to the numerical method used to solve the dynamical system of equations. Thus, one identifies the kind of blocks to be used, such as: comparators, shifters, adders, subtractors, multipliers, multiplexers, and so on.

The chaotic oscillators mentioned above can be implemented with similar hardware blocks, except the PWL functions, which can be implemented including comparators to solve conditionals like in (4.27). Each block having ports for processing words of length = 28-bits. To control the speed of the hardware, all blocks used herein have pins to include an asynchronous reset and the clock signal. Figure 4.11 shows the generic blocks for the adder, multiplier and subtractor, more detailed descriptions with other formats can be seen in [25, 33].

Among the currently available VHDL software resources, we list the VHDL code for 4 blocks: adder, multiplier, subtractor and comparator. All of them are for processing 28 bits. The first block is the adder that includes clock and reset. In this case, using IEEE library, the entity is named "sumador," having 5 ports for allocating clock, reset and 3 data buses. The architecture is named "complicado," it uses reset and clock signals to process data within the ports of entity "sumador," as shown below. The VHDL code for the subtractor is quite similar to the one for the adder, juts by changing plus by minus.

```
1 library ieee;
2 use ieee.std_logic_1164.all;
3 use ieee.numeric_std.all;
```

```
entity sumador is
port(

    clock_50: in std_logic;
    reset: in std_logic;
    dato_1: in std_logic_vector(27 downto 0);
    dato_2: in std_logic_vector(27 downto 0);
    dato_3: out std_logic_vector(27 downto 0):=(others => '0')
);
end sumador;
architecture complicado of sumador is
begin
    process(clock_50,reset,dato_1,dato_2)
    begin
        if reset = '1' then
            dato_3 <= (others => '0');
        elsif rising_edge(clock_50) then
            dato_3 <= std_logic_vector(signed(dato_1) + signed(dato_2));
        end if;
    end process;
end complicado;
```

**Listing 4.1** VHDL code for an adder

```
library ieee;
use ieee.std_logic_1164.all;
use ieee.numeric_std.all;
entity restador is
port(

    clock_50: in std_logic;
    reset: in std_logic;
    dato_1: in std_logic_vector(27 downto 0);
    dato_2: in std_logic_vector(27 downto 0);
    dato_3: out std_logic_vector(27 downto 0):="0000000000000000000000000000"
);
end restador;
architecture complicado of restador is
begin
    process(clock_50,reset,dato_1,dato_2)
    begin
        if reset = '1' then
            dato_3 <= (others => '0');
        elsif rising_edge(clock_50) then
            dato_3 <= std_logic_vector(signed(dato_1) - signed(dato_2));
        end if;
    end process;
end complicado;
```

**Listing 4.2** VHDL code for a subtractor

For the case of the multiplier, when multiplying two 28-bits words, one should identify the bits allocating the result, so that a more elaborated architecture is created. See the following VHDL code where the result is assigned to dato_3 but prior to this, the multiplication is performed between two data and assigned to an extra variable named "sena1," from which the bits allocating the result are identified from 51 down to 24.

```
library ieee;
use ieee.std_logic_1164.all;
use ieee.numeric_std.all;
entity multiplicador is
    port(
```

```
6     clock_50: in std_logic;
7     reset: in std_logic;
8     dato_1: in std_logic_vector(27 downto 0);
9     dato_2: in std_logic_vector(27 downto 0);
10    dato_3: out std_logic_vector(27 downto 0):=(others=>'0')
11 );
12 end multiplicador;
13 architecture complicado of multiplicador is
14 signal senal: signed(55 downto 0):= (others=>'0');
15 begin
16    process(clock_50,reset)
17    begin
18        if reset = '1' then
19            dato_3 <= (others => '0');
20        elsif rising_edge(clock_50) then
21            senal <= (signed(dato_1)*signed(dato_2));
22            dato_3 <= std_logic_vector(senal(51 downto 24));
23        end if;
24    end process;
25 end complicado;
```

**Listing 4.3** VHDL code for a multiplier

The comparator is a more complex block. If one thinks on implementing PWL functions, as the ones for the saturated function series, one should declare as many ports as saturation levels and slopes the PWL function has. The architecture is also more elaborated and real constants (as the break-points) can be declared by binary notation, as shown in the following VHDL code.

```
1  library ieee;
2  use ieee.std_logic_1164.all;
3  use ieee.std_logic_unsigned.all;
4  use ieee.std_logic_arith.all;
5
6  entity comparador is
7     port(
8     clock_50: in std_logic;
9     reset: in std_logic;
10    dato_sat0: in std_logic_vector(27 downto 0);          —Saturation Level
11    .
12    .
13    .
14    dato_sat(N-1): in std_logic_vector(27 downto 0);   —Saturation Level
15    dato_pen0: in std_logic_vector(27 downto 0);           —Slope
16    .
17    .
18    .
19    dato_pen(N-2): in std_logic_vector(27 downto 0); —Slope
20    dato_X: in std_logic_vector(27 downto 0);              — X
21    dato_S: out std_logic_vector(27 downto 0)              —Output
22    );
23 end comparador;
24
25 architecture complicado of comparador is
26 constant q0: std_logic_vector(27 downto 0):="1000000110011001100110011010";
27          —Breakpoints in Binary
28 .
29 .
30 .
31 constant q(2*N-1): std_logic_vector(27 downto 0):="0111111001100110011001100110
32       ";  —Breakpoints in Binary
33 begin
34    process(clock_50,reset,dato_X)
35    begin
```

```
36         if reset='1' then
37                 dato_S<= (others => '0');
38         elsif rising_edge(clock_50) then
39                 if dato_X > q0 AND dato_X < q1 then
40                     dato_S <= dato_sat0;
41                 .
42                 .
43                 .
44                 elsif dato_X > q(2*N-2) AND dato_X < q(2*N-1) then
45                     dato_S <= dato_sat1;
46                 else
47                     dato_S <= dato_pen0;
48                 end if;
49         end if;
50     end process;
51 end complicado;
```

**Listing 4.4** VHDL code for a comparator implementing a PWL function based on saturated function series

The blocks described above can be used to implement the chaotic oscillator based on saturated function series. The comparator will augment the number of ports in a proportional fashion as the number of scrolls increase, as mentioned above.

As shown in [25], a mathematical model can be solved by applying different numerical methods. When using the Forward Euler one, the discretized equations for the chaos generator based on saturated function series are given by (4.32), where initial conditions are required and which can be set using multiplexers [25, 33]. In this case, one needs to control a loop for performing the iterations, as shown by

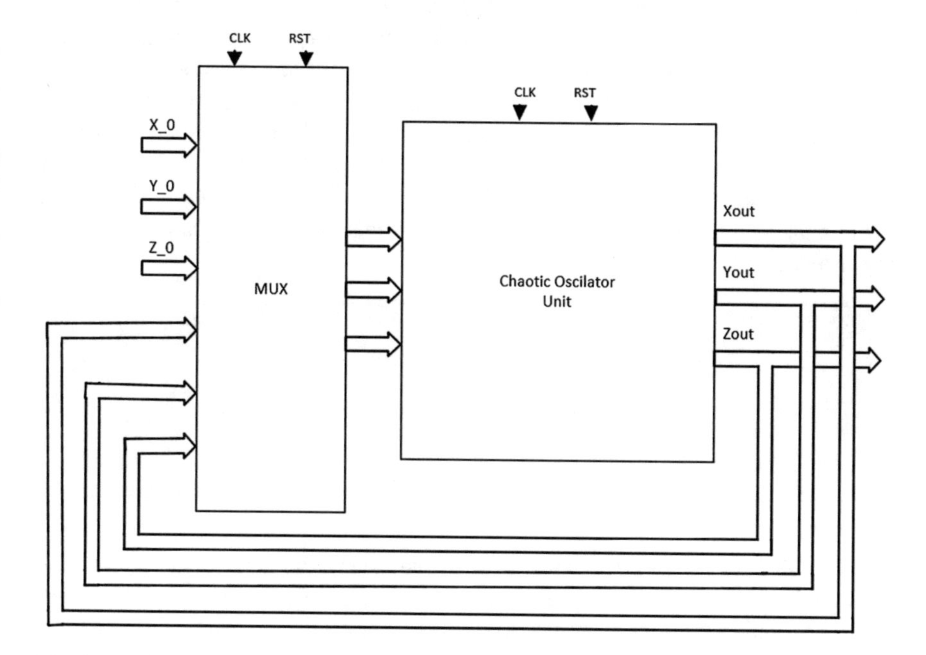


**Fig. 4.12** Block description of (4.32), showing the state variables $x$, $y$ and $z$

Fig. 4.12. Where the chaotic oscillator unit embeds the hardware realization of (4.32), in which one can identify the required blocks for implementing the chaos generator using FPGAs. Sign plus requires and which is an adder, minus a subtractor, and those state variables multiplied by the step size $h$, require a multiplier, also $d_1$ is multiplying the PWL function that can be implemented using comparators.

$$\begin{aligned} x_1 &= x_0 + hy_0 \\ y_1 &= y_0 + hz_0 \\ z_1 &= z_0 + h(-ax - by - cz + d_1 f(x; q)) \end{aligned} \tag{4.32}$$

Equation (4.32) can be described by VHDL code, while initial conditions are included using a multiplexer and clock control that allow processing $x_0$, $y_0$ and $z_0$ just to begin the iterations.

## 4.4 VHDL Code Generation

VHDL code for chaos generators can be automatically generated as described below. The PWL functions can be implemented by comparator blocks that increase in a systematic way as a function of the number of scrolls being generated. For instance, again considering the chaotic oscillator based on the PWL function for saturated series, it has four coefficients $a$, $b$, $c$, and $d_1$, and according to the number of scrolls, one knows the number of break-point and saturation levels. Using Python, the following pseudocode performs the automatic synthesis of chaos generators based on saturated function series as PWL function. As one sees, the input parameters are the four coefficients of the dynamical system, and then the PWL function requires the number of scrolls, break-points $q_i$ and saturation levels $k_i$. Using these values one can compute the slopes $m_i$ of the PWL function (see Fig. 4.7). The for cycle is simulating the dynamical system with a step size (step) estimated from the evaluation of the eigenvalues.

```
1 Enter data (Coefficients, Scrolls, Breakponits, Saturation Level)
2   for(i = 0 to simulation time)
3     x(after) = x+step f(x)
4     y(after) = y+step f(y)
5     z(after) = z+step f(z, Scrolls, Breakpoints, Sturation Level)
6   end for
7 print(graphics)
```

**Listing 4.5** Pseudocode for the automatic synthesis of chaos generators based on the saturated function series as PWL function

Using Python one can verify that using these values, the dynamical system consisting of the three state variables $x$, $y$, and $z$, generate chaotic behavior. It can be observed plotting the data between two state variables, e.g., $x$–$y$ to observe an attractor. After this, one can provide data in binary format and assign the corresponding binary words to the blocks: multiplier, adder, subtractor, comparator, and multiplexer.

**Table 4.2** PWL description for generating symmetric scrolls

| Scrolls | Saturation level | Break-point |
|---|---|---|
| 2 | −1, 1 | −0.0165, 0.0165 |
| 3 | −2, 0, 2 | −1.0165, −0.9835, 0.9835, 1.0165 |
| 5 | −4, −2, 0, 2, 4 | −3.0165, −2.9825, −1.0165, −0.9835, 0.9835, 1.0165, 2.9835, 3.0165 |
| 6 | −5, −3, −1, 1, 3, 5 | −4.0165, −3.9835, −2.0165, −1.9835, −0.0165, 0.0165, 1.9835, 2.0165, 3.9835, 4.0165 |

**Table 4.3** PWL description for generating nonsymmetric scrolls

| Scrolls | Saturation level | Break-point |
|---|---|---|
| 2 | −1, 2 | −0.0165, 0.0165 |
| 3 | −1.5, 0, 1.7 | −1.0165, −0.9835, 0.9835, 1.0165 |
| 5 | −5.1, −2.03, 0, 2, 4.96 | −3.0165, −2.9825, −1.0165, −0.9835, 0.9835, 1.0165, 2.9835, 3.0165 |
| 6 | −6.07, −3, −1.03, 1.05, 3, 5.95 | −4.0165, −3.9835, −2.0165, −1.9835, −0.0165, 0.0165, 1.9835, 2.0165, 3.9835, 4.0165 |

The identification of such blocks is done from the discretized equations. For example: from (4.32), signal $x_1$ requires a multiplier to evaluate $hy_0$ and an adder to sum that with $x_0$. Signal $y_1$ also requires a multiplier and an adder. However, signal $z_1$ requires 5 multipliers, 3 adders, 1 subtractor, and a comparator to implement the PWL function.

Tables 4.2 and 4.3 list the values for the VHDL code generation of chaos generators having from 2 to 6-scrolls, with symmetric and nonsymmetric PWL functions, respectively. It is appreciated that the break-points and saturated levels are nonsymmetric in the second case, so that the computed slopes change in symmetric and nonsymmetric PWL functions.

The VHDL code can be automatically generated by identifying the way the PWL function is implemented. After an analysis of the chaos generators based on the PWL functions consisting on saturated functions series, sawtooth and negative slopes (Chua's circuit), described at Sect. 4.1, one can conclude that the hardware increases in a proportional fashion with respect to the number of scrolls. For example, from the chaotic oscillator based on saturated PWL functions, for generating 2-scrolls the required hardware are 12 multipliers, 8 adders, and 2 subtractors. Each additional scroll requires 5 multipliers, 3 adders and 1 subtractor, so that this can be expressed by (4.33), where $n$ indicates the number of scrolls. On the other hand, the PWL function can be augmented from (4.34), which is for generating from 2-scrolls to $n$.

$$\begin{aligned} Multipliers &= 5(n-2)+12 \\ Adders &= 3(n-2)+8 \\ Subtractors &= n \end{aligned} \tag{4.33}$$

$$
\begin{aligned}
&if\,(x_1 > q_l \; and \; x_1 < q_{l+1})\\
&\quad f_0 = k_p\\
&while \; l < (2n-2)\\
&\quad if\,(x_1 >= q_{l+1} \; and \; x1 <= q_{l+2})\\
&\quad\quad f_1 = ((k_{p+1} - k_p)/(q_{l+2} - q_{l+1}))\\
&\quad\quad f_2 = (x_1 - ((q_{l+1} + q_{l+2})/2)) + ((k_{p+1} + k_p)/2)\\
&\quad\quad f_0 = f_1 f_2\\
&\quad else - if\,(x_1 > q_{l+2} \; and \; x_1 < q_{l+3})\\
&\quad\quad f_0 = k_{p+1}\\
&\quad l = l + 2\\
&\quad p = p + 1
\end{aligned}
\tag{4.34}
$$

Simulation results for the symmetric attractors are shown in Figs. 4.13, 4.14, 4.15 and 4.16, and for nonsymmetric attractors they are shown in Figs. 4.17, 4.18, 4.19 and 4.20. FPGA realizations of these chaos generators lead us to the experimental results shown in Figs. 4.21, 4.22, 4.23 and 4.24. On the other hand, as the dynamical system has 3 state variables, 3 Lyapunov exponents can be computed from which one must be maximum (MLE) to guarantee chaotic regime [32]. Table 4.4 lists the computed MLEs for generating 2, 3, 5, and 6-scrolls with symmetric and nonsymmetric PWL functions, where it can be appreciated what is well known: MLE increases as the number of scrolls being augmented. In addition, better MLE values can be obtained when optimizing it as already shown in [32, 34].

## 4.5 Bifurcation Diagrams

A dynamical system can change its behavior as its parameters vary, such changes can be appreciated by evaluating the bifurcation diagram [35, 36]. Considering again the chaos generator based on the saturated PWL function, varying the coefficients $a$, $b$, $c$, and $d_1$ leads to compute better values for MLE. Table 4.5 shows the computed coefficients and their associated values for MLE, entropy and fractal dimension, the initial conditions to the three state variables $x$, $y$, and $z$, were set to $[0.1, 0, 0]$.

For the chaos generators listed in Table 4.5, the bifurcation diagrams are generated [34], as shown in Figs. 4.25, 4.26 and 4.27, where $c$ was selected as bifurcation parameter because it is the more sensitive one.

## 4.6 Multi-scroll Chaotic Attractors with High MLE and Entropy

In the recent literature one can find works discussing the modeling, simulation, and circuit realization of different kinds of continuous-time multi-scroll chaotic attractors. However, very few works describe the experimental realization of attractors

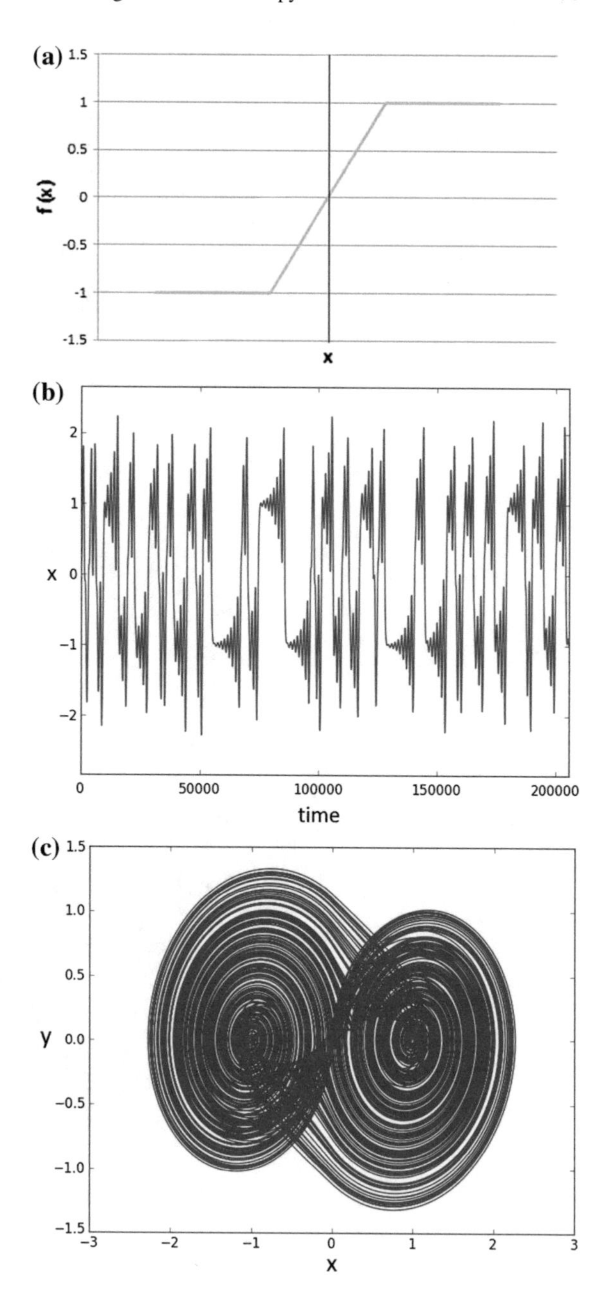


**Fig. 4.13** Simulation results for the symmetric and 2-scrolls attractor. **a** PWL function. **b** State variable $x$. **c** Portrait $x$–$y$

having high maximum Lyapunov exponent (MLE) and high entropy, which are desirable characteristics to guarantee better chaotic unpredictability. For instance, two chaotic oscillators having the same MLE values can behave in a very different way, e.g., showing different entropy values. That way, in this section we list

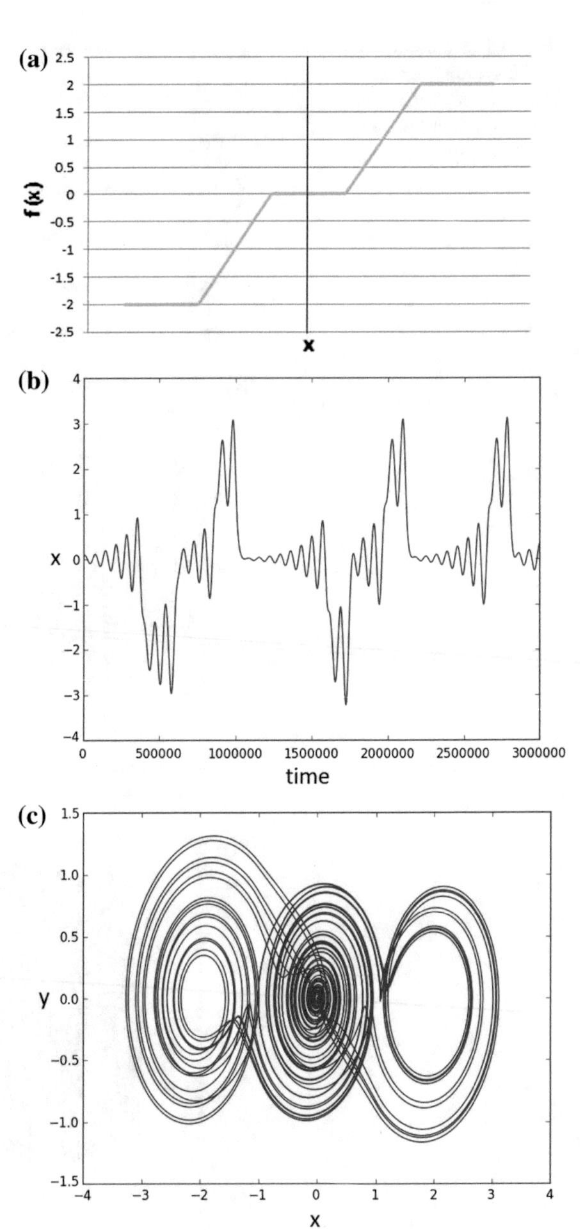


**Fig. 4.14** Simulation results for the symmetric and 3-scrolls attractor. **a** PWL function. **b** State variable $x$. **c** Portrait $x$–$y$

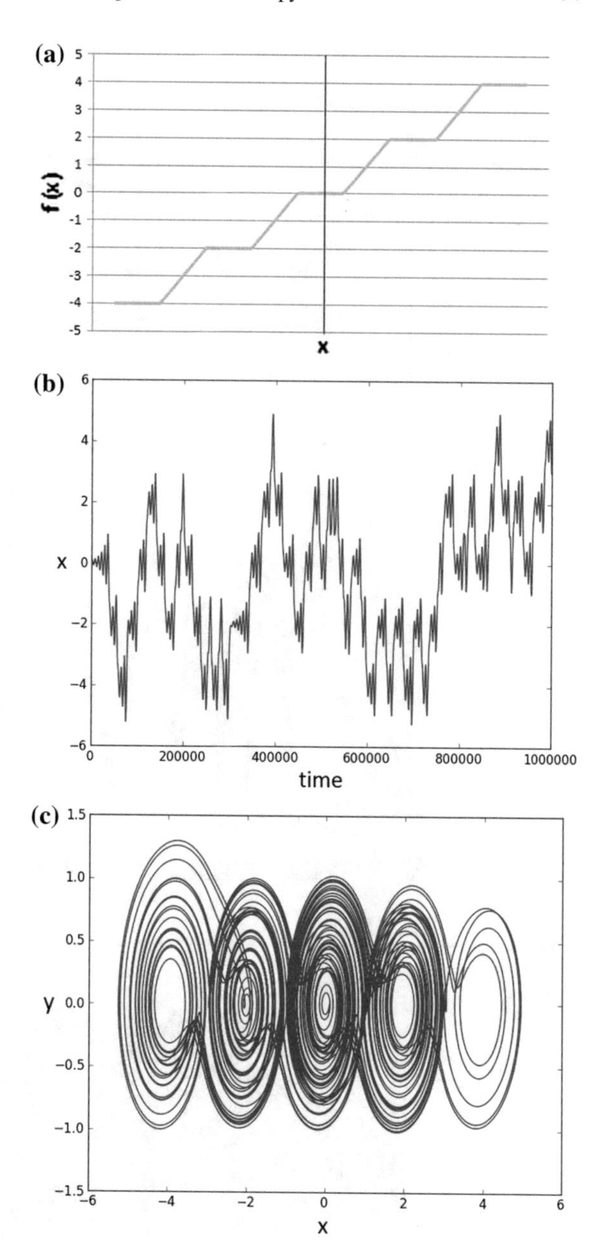


**Fig. 4.15** Simulation results for the symmetric and 5-scrolls attractor. **a** PWL function. **b** State variable $x$. **c** Portrait $x$–$y$

some values of an optimized multi-scroll chaotic oscillator with both high MLE and entropy [37]. The optimization algorithm proposed in [32], is applied herein. It is based on the evolutionary algorithm known as nondominated sorting genetic algorithm (NSGA-II) [38], and it optimizes two characteristics, then: a bi-objective

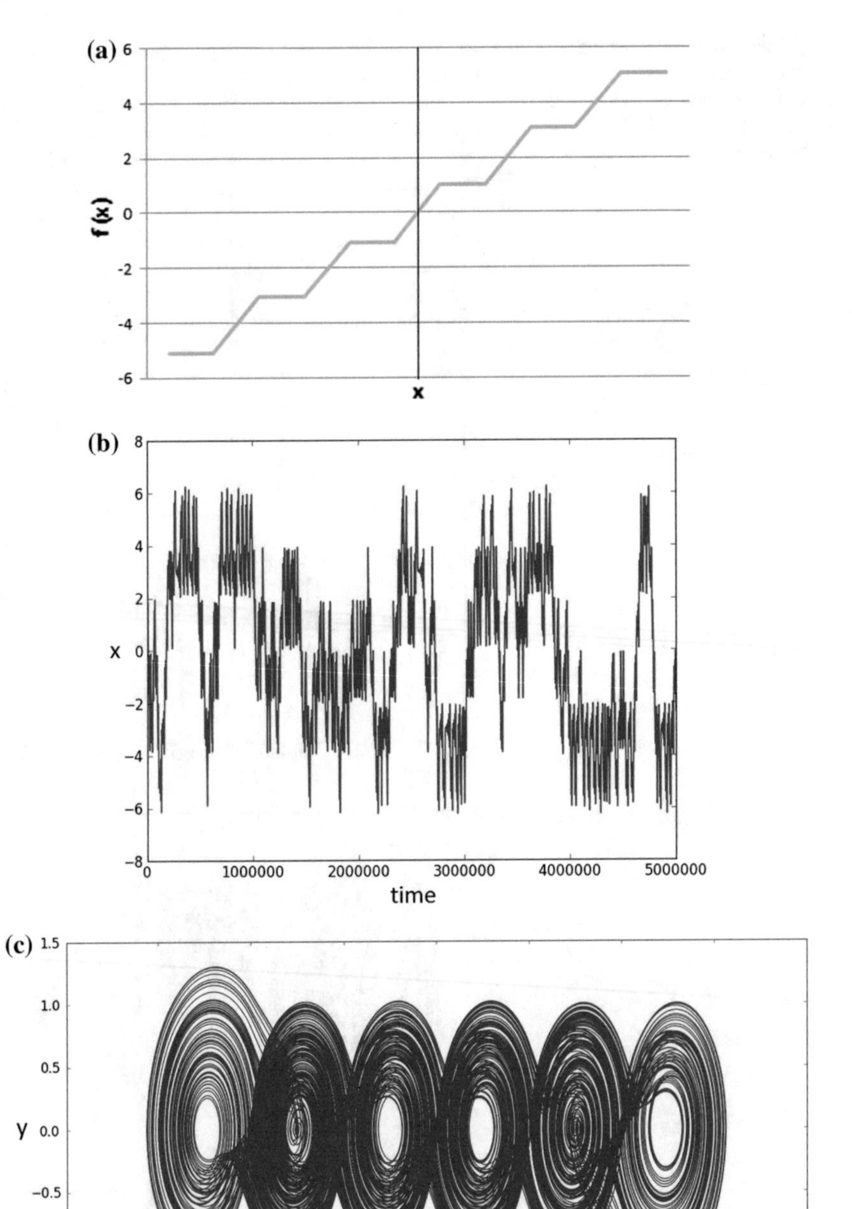


**Fig. 4.16** Simulation results for the symmetric and 6-scrolls attractor. **a** PWL function. **b** State variable $x$. **c** Portrait $x$–$y$

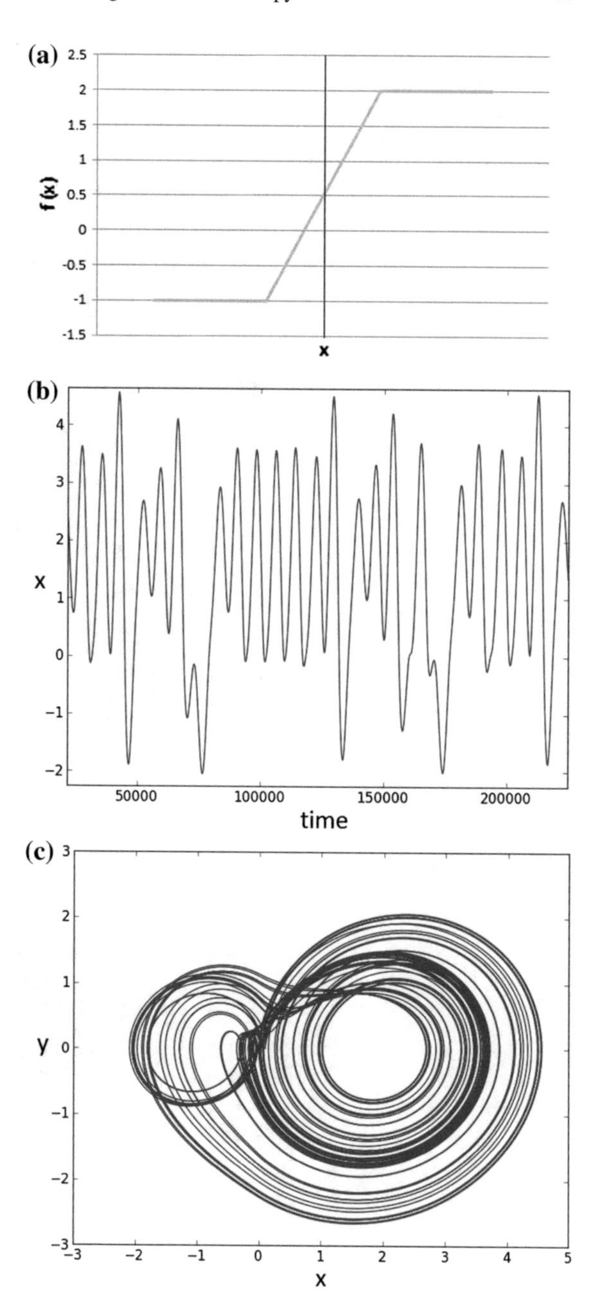


**Fig. 4.17** Simulation results for the nonsymmetric and 2-scrolls attractor. **a** PWL function. **b** State variable $x$. **c** Portrait $x$–$y$

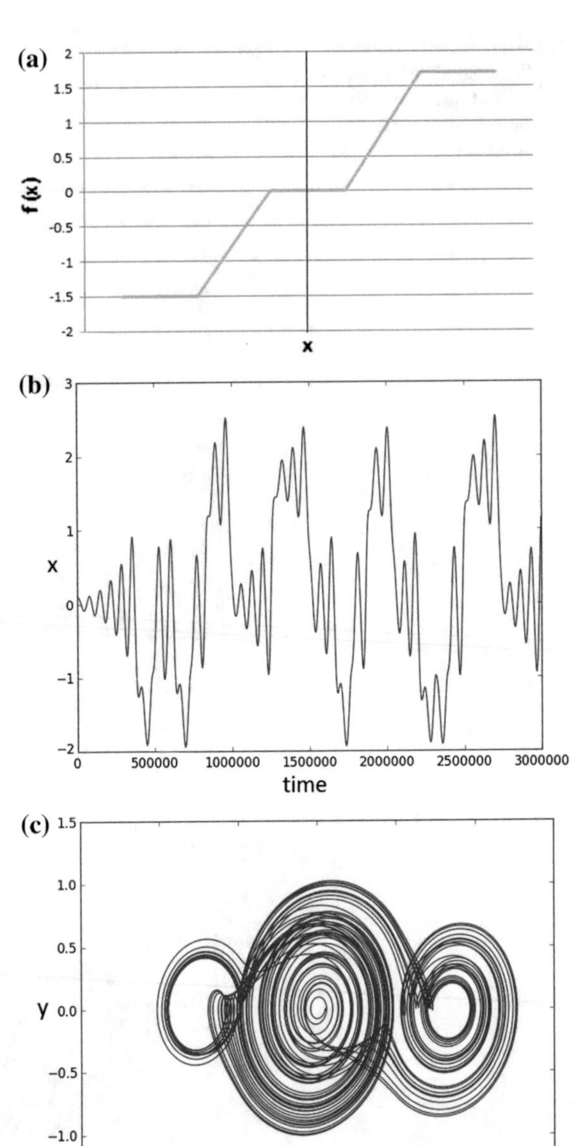


**Fig. 4.18** Simulation results for the nonsymmetric and 3-scrolls attractor. **a** PWL function. **b** State variable $x$. **c** Portrait $x$–$y$

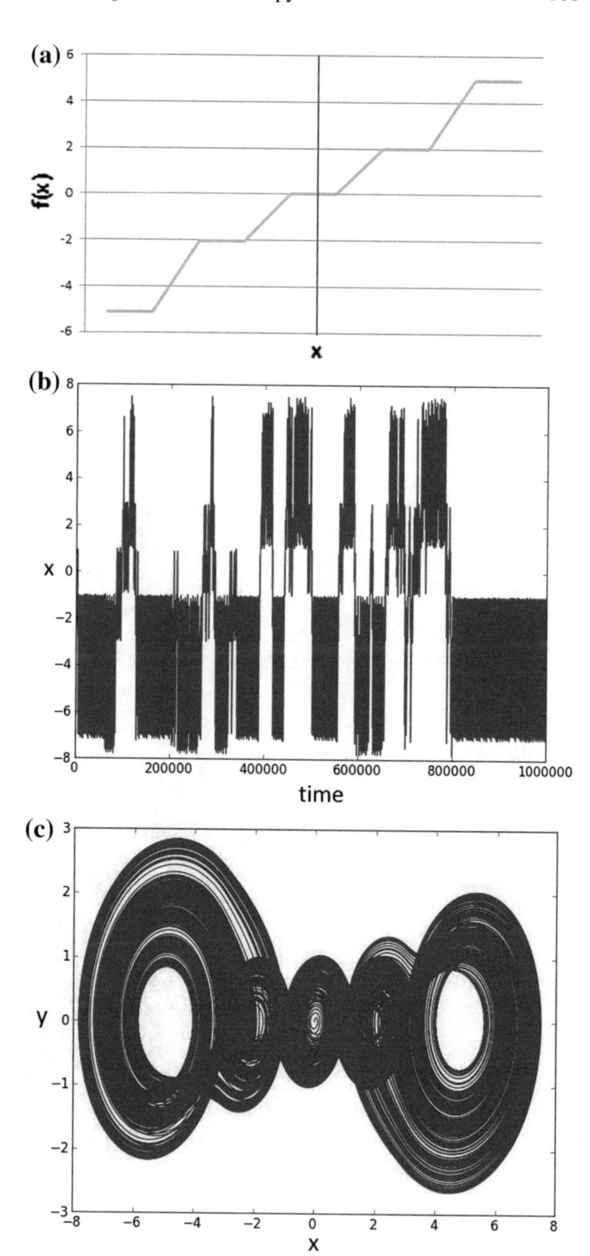


**Fig. 4.19** Simulation results for the nonsymmetric and 5-scrolls attractor. **a** PWL function. **b** State variable $x$. **c** Portrait $x$–$y$

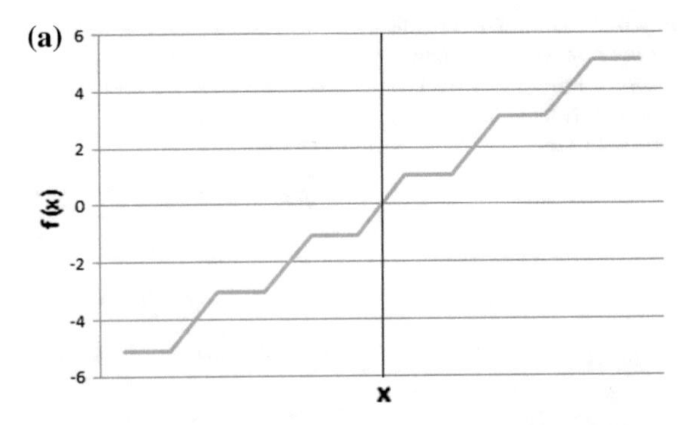


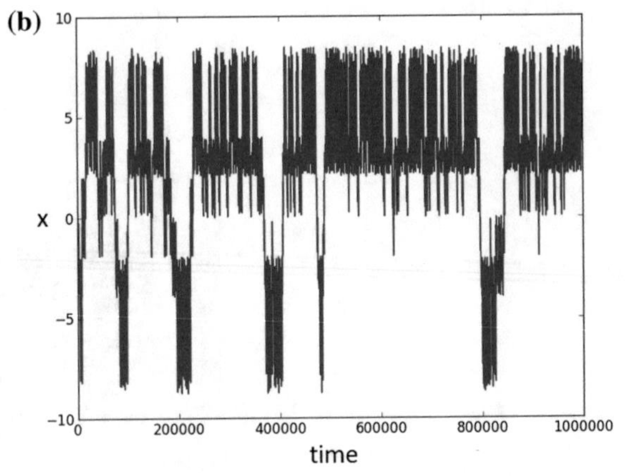


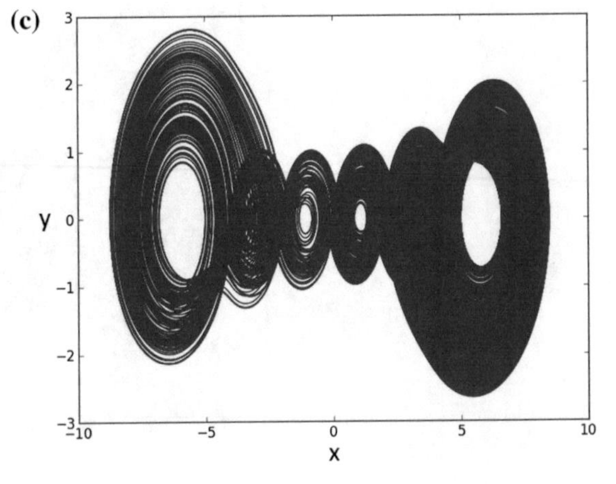


**Fig. 4.20** Simulation results for the nonsymmetric and 6-scrolls attractor. **a** PWL function. **b** State variable $x$. **c** Portrait $x$–$y$

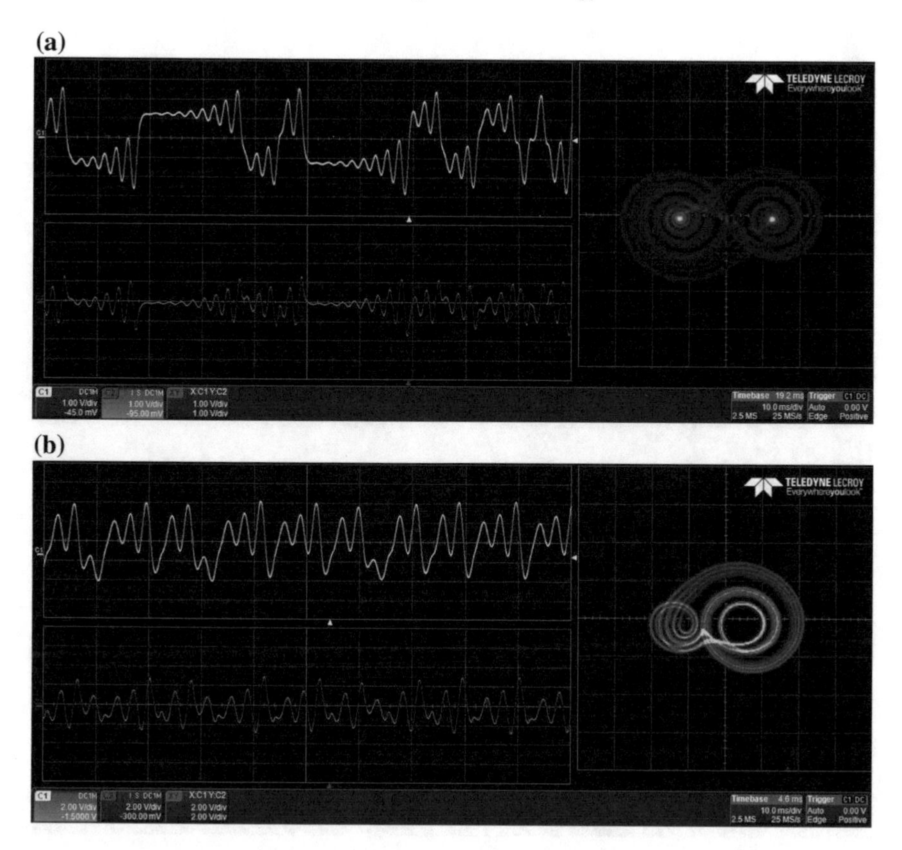

Fig. 4.21 Experimental results for generating 2-scrolls with symmetric and nonsymmetric PWL functions. **a** Symmetric. **b** Nonsymmetric

optimization problem is encoded: (i) to maximize MLE, and (ii) to minimize the variability in the oscillator's phase-space transitions or the trajectories.

### *4.6.1 Lyapunov Exponents*

Lyapunov exponents are asymptotic measures that characterize the average rate of growth (or shrinking) of small perturbations to the solutions of a dynamical system. They provide quantitative measures of response sensitivity of a dynamical system to small changes in initial conditions. In a continuous time dynamical system modeled by ordinary differential equations, the number of Lyapunov exponents is equal to the number of states variables in the dynamical system, so that at least three state variables are required to generate chaotic behavior. This section verifies what is already known

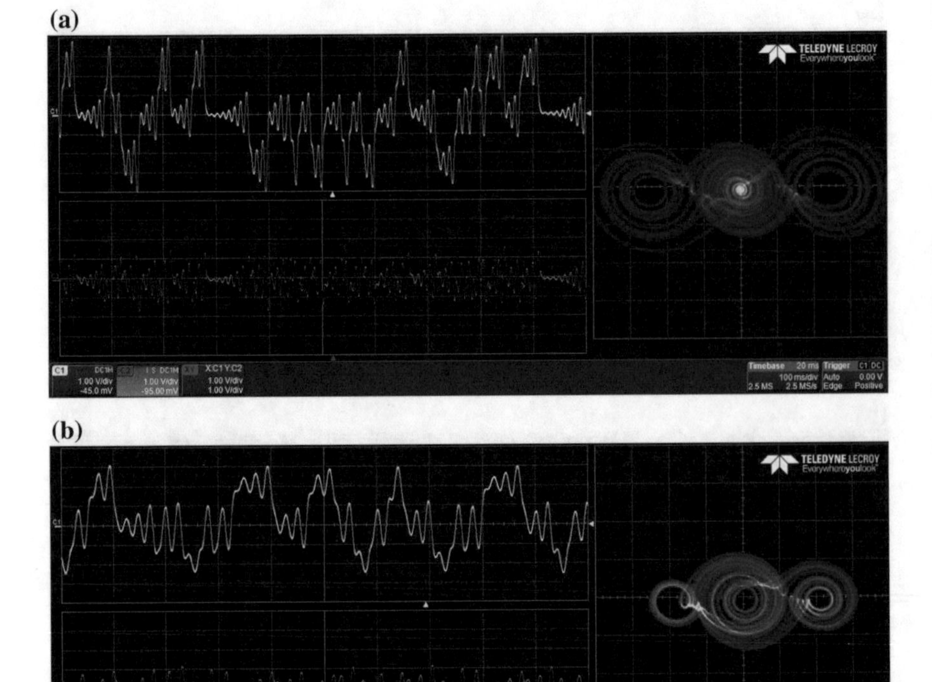


**Fig. 4.22** Experimental results for generating 3-scrolls with symmetric and nonsymmetric PWL functions. **a** Symmetric. **b** Nonsymmetric

that by increasing the number of scrolls both the MLE and its associated entropy increase in a similar proportion [37].

Lets us consider an $n$-dimensional dynamical system:

$$\dot{x} = f(x), \quad x \in \mathbb{R}^n, \quad t > 0, \quad x(0) = x_0 \tag{4.35}$$

where $x$ and $f$ are $n$-dimensional vector fields. To determine the $n$-Lyapunov exponents of the system one have to find the long-term evolution of small perturbations to a trajectory, which are determined by the variational equation of (4.35),

$$\dot{y} = \frac{\partial f}{\partial x}\big(x(t)\big)y = J\big(x(t)\big)y \tag{4.36}$$

where $J$ is the $n \times n$ Jacobian matrix of $f$. A solution to (4.36) with a given initial perturbation $y(0)$ can be written as

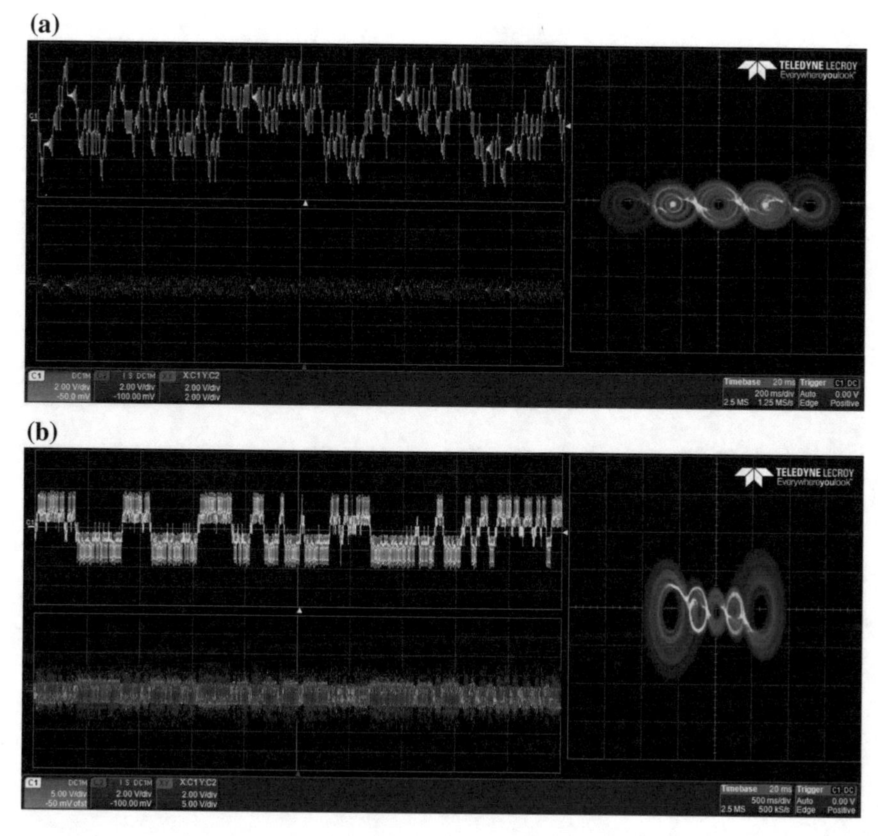


**Fig. 4.23** Experimental results for generating 5-scrolls with symmetric and nonsymmetric PWL functions. **a** Symmetric. **b** Nonsymmetric

$$y(t) = Y(t)y(0) \tag{4.37}$$

with $Y(t)$ as the fundamental solution satisfying

$$\dot{Y} = J\big(x(t)\big)Y, \quad Y(0) = I_n \tag{4.38}$$

Here $I_n$ denotes the $n \times n$ identity matrix. If one considers the evolution of an infinitesimal $n$-parallelepiped $[p_1(t), \ldots, p_n(t)]$ with the axis $p_i(t) = Y(t)p_i(0)$ for $i = 1, \ldots, n$, where $p_i(0)$ denotes an orthogonal basis of $\mathbb{R}^n$. The $i$th Lyapunov exponent, which measures the long-time sensitivity of the flow $x(t)$ with respect to the initial data $x(0)$ at the direction $p_i(t)$, is defined by the expansion rate of the length of the $i$th axis $p_i(t)$ and is given by

$$\lambda_i = \lim_{t \to \infty} \frac{1}{t} \ln \left\| p_i(t) \right\| \tag{4.39}$$

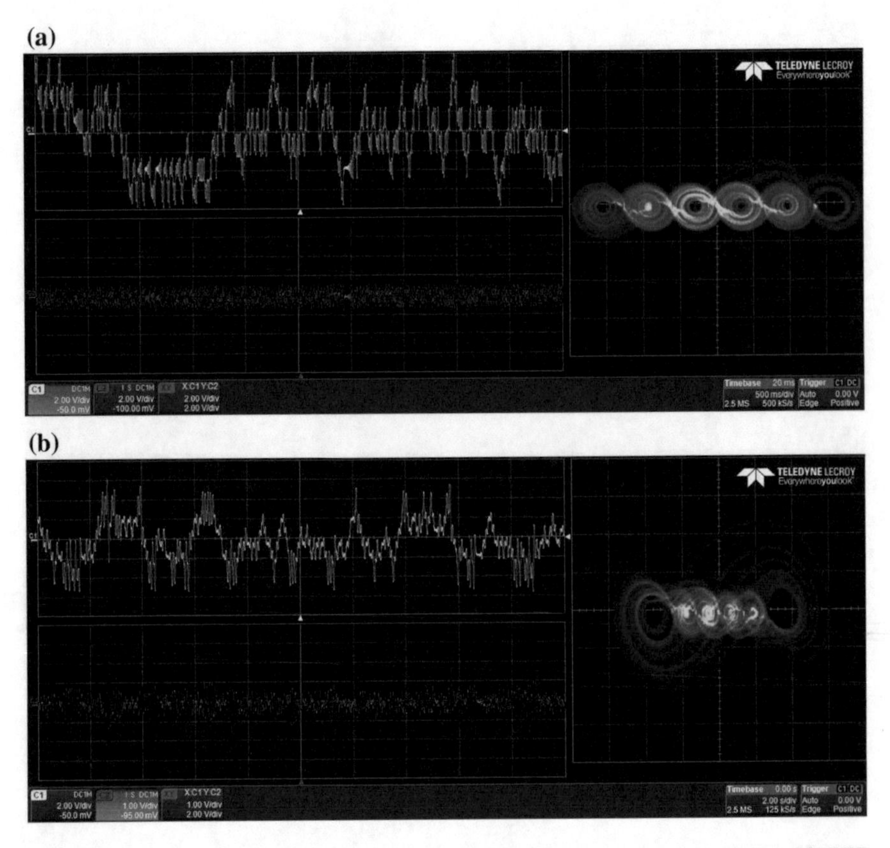


**Fig. 4.24** Experimental results for generating 6-scrolls with symmetric and nonsymmetric PWL functions. **a** Symmetric. **b** Nonsymmetric

In summary, the Lyapunov exponents can be computed as follows [20, 32, 39, 40]:

1. Initial conditions of the system and the variational system are set to $X_0$ and $I_{n\times n}$, respectively.
2. The systems are integrated by several steps until an orthonormalization period TO is reached. The integration of the variational system $Y = [y_1, y_2, y_3]$ depends on the specific Jacobian that the original system $X$ is using in the current step.
3. The variational system is orthonormalized using the standard Gram–Schmidt method [41], and the logarithm of the norm of each Lyapunov vector contained in $Y$ is obtained and accumulated in time.
4. The next integration is carried out using the new orthonormalized vectors as initial conditions. This process is repeated until the full integration period $T$ is reached.
5. The Lyapunov exponents are obtained by

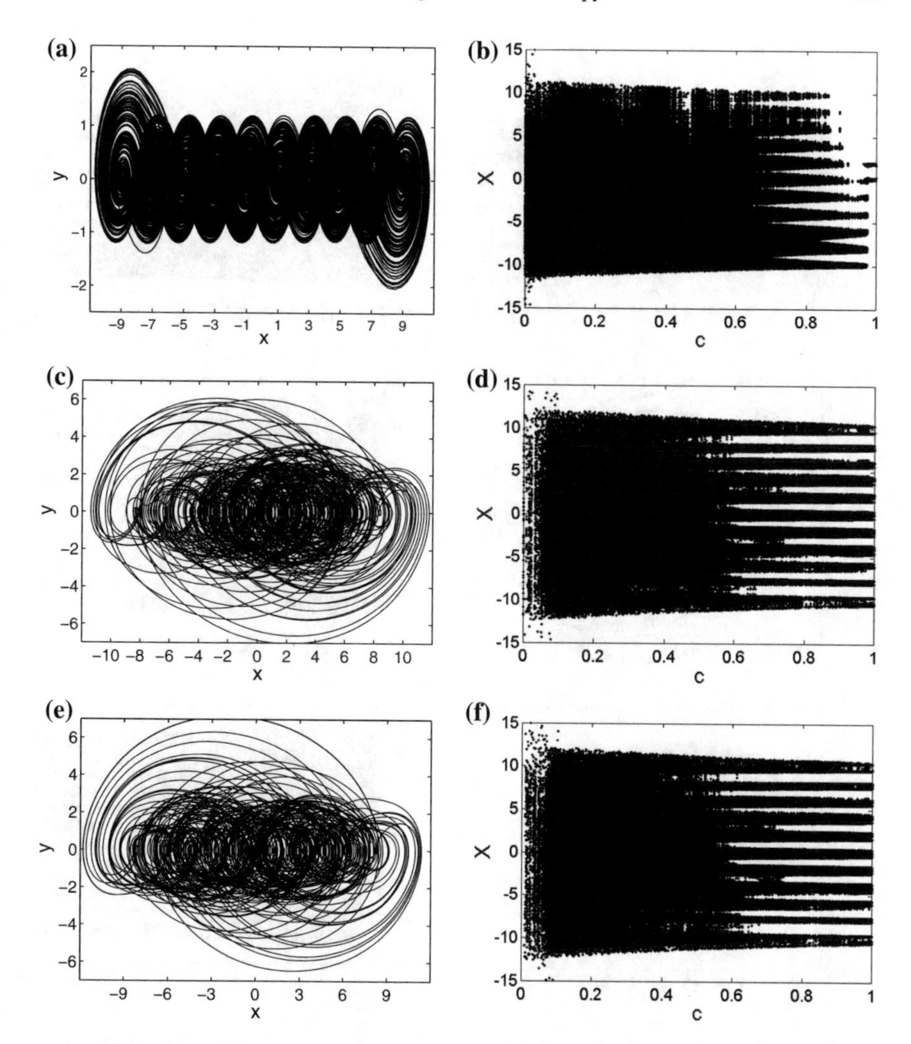


**Fig. 4.25** Bifurcation diagrams and portraits for generating 10-scrolls with different MLE from Table 4.5. **a** MLE = 0.165950. **b** MLE = 0.165950. **c** MLE = 0.775137. **d** MLE = 0.775137. **e** MLE = 0.776849. **f** MLE = 0.776849

$$\lambda_i \approx \frac{1}{T} \sum_{j=TO}^{T} \ln \|y_i\|$$

The time-step selection was set as in [40], using the minimum absolute value of all the eigenvalues of the system $\lambda_{min}$, and $\psi$ was chosen well above the sample theorem as 50.

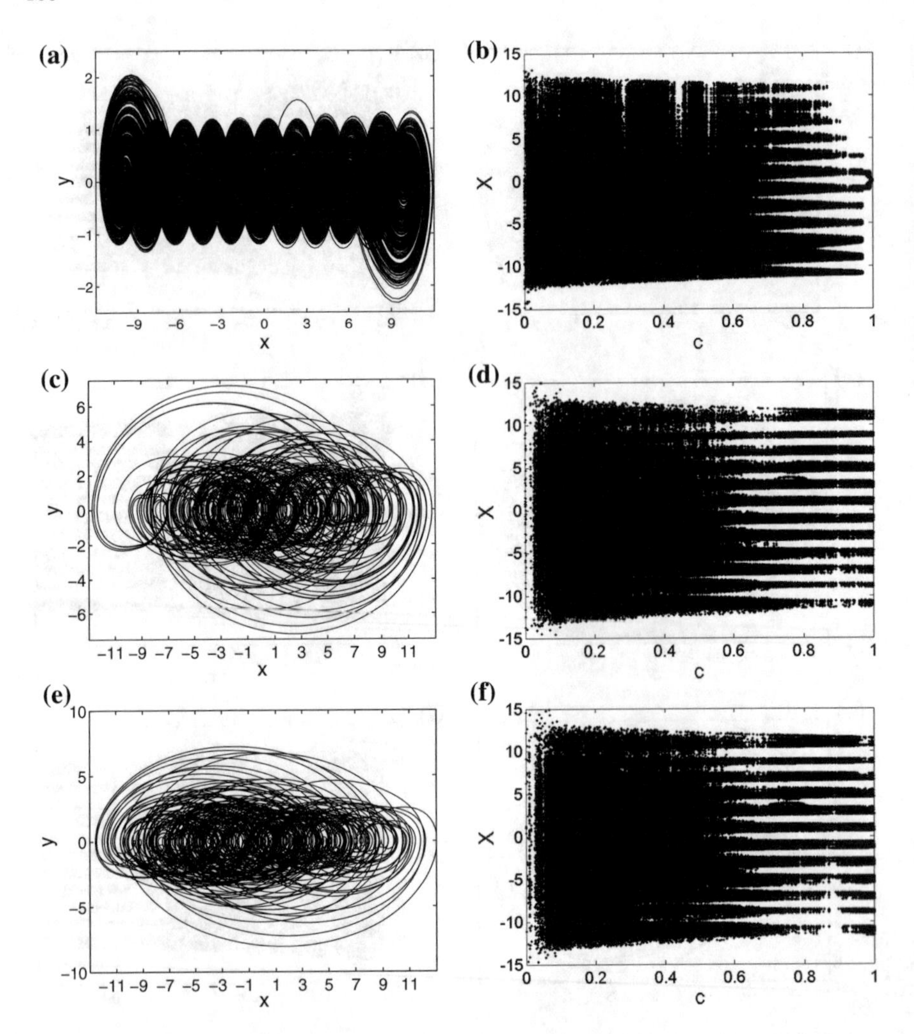


**Fig. 4.26** Bifurcation diagrams and portraits for generating 11-scrolls with different MLE from Table 4.5. **a** MLE = 0.157250. **b** MLE = 0.157250. **c** MLE = 0.772160. **d** MLE = 0.772160. **e** MLE = 0.785336. **f** MLE = 0.785336

$$t_{\text{step}} = \frac{1}{\lambda_{\min}\psi}$$

The orthogonalization period *TO* was chosen about 50 $t_{step}$. This procedure is used herein as in [32] to optimize the MLE.

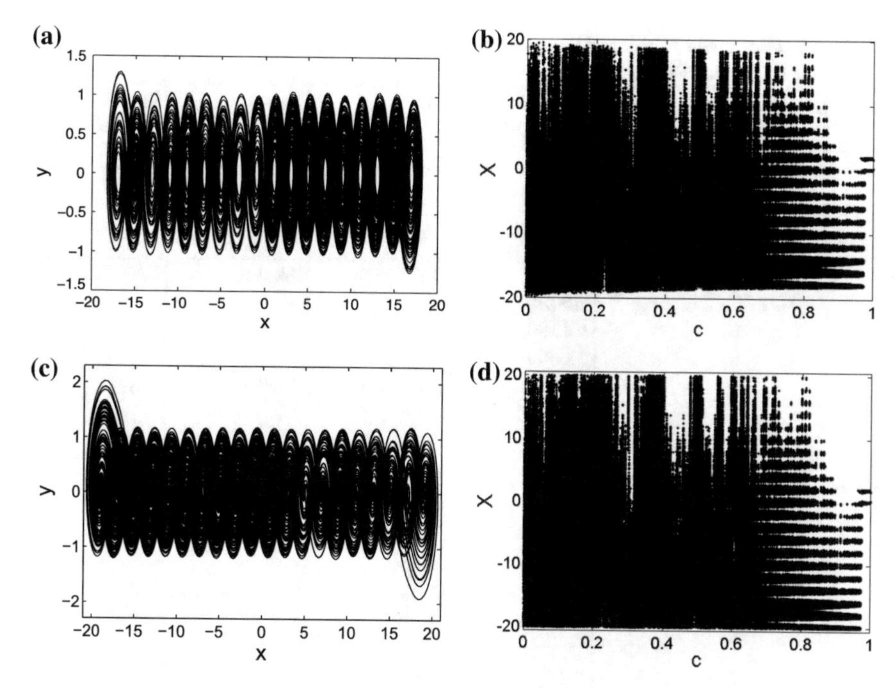


**Fig. 4.27** Bifurcation diagrams and portraits for generating 18 and 20-scrolls. **a** MLE = 0.1813334. **b** MLE = 0.1813334. **c** MLE = 0.1813339. **d** MLE = 0.1813339

**Table 4.4** Computed MLE values

| Scrolls | MLE for symmetric PWL | MLE for nonsymmetric PWL |
|---|---|---|
| 2 | 0.085481 | 0.087523 |
| 3 | 0.088219 | 0.080768 |
| 5 | 0.106306 | 0.106320 |
| 6 | 0.111062 | 0.111581 |

### 4.6.2 *Evaluation of Entropy*

For chaotic oscillators, the entropy is an alternative choice to Lyapunov exponents because it reveals aspects of the underlying dynamical system (i.e., it quantifies the stretching and the folding aspects at the same time). The entropy rates of growth are an interesting parameter to quantify disorder in chaotic oscillators. In the same direction, as chaotic attractors can be recognized by visual inspection in their phase-space portraits, a numerical quantification of chaos is performed by optimizing MLE. The entropy has also some relationships of interest as for the sum of Lyapunov exponents [39, 42], which measure the instability of nearby trajectories. The entropy

**Table 4.5** Coefficient values and their associated MLE, entropy and fractal dimension for generating 10, 11, 18, and 20-scrolls

| Scrolls | PWL segments | Coefficient $(a, b, c, d_1)$ | MLE | Entropy | Fractal dimension |
|---|---|---|---|---|---|
| 10 | 19 | 0.7000, 0.7000, 0.7000, 0.7000 | 0.16595 | 2.6869 | 2.1919 |
| | | 1.0000, 0.5160, 0.1190, 1.0000 | 0.775137 | 2.8490 | 2.8743 |
| | | 1.0000, 0.5130, 0.1180, 1.0000 | 0.776849 | 2.8628 | 2.8755 |
| 11 | 21 | 0.7000, 0.7000, 0.7000, 0.7000 | 0.15725 | 2.7460 | 2.1832 |
| | | 1.0000, 0.4820, 0.1100, 1.0000 | 0.77216 | 2.9148 | 2.8750 |
| | | 1.0000, 0.4850, 0.0930, 1.0000 | 0.785336 | 2.9950 | 2.8968 |
| 18 | 35 | 0.7000, 0.7000, 0.7000, 0.7000 | 0.1813334 | 2.9003 | 2.20873 |
| 20 | 39 | 0.7000, 0.7000, 0.7000, 0.7000 | 0.1813339 | 3.0155 | 2.90035 |

is computed herein by applying the algorithm presented by Moddemeijer, which is online available at http://www.cs.rug.nl/~rudy/matlab/.

Some values (cases) of the coefficients $a$, $b$, $c$, and $d_1$, associated to the optimized MLEs for different number of scrolls, and their corresponding entropies are listed in Tables 4.6, 4.7 and 4.8. These tables clearly show that by increasing the number of scrolls, when the chaotic oscillator is optimized both MLE and the entropy increase.

**Table 4.6** Optimized MLE and its associated entropy for generating 2-scrolls

| Case | $a$ | $b$ | $c$ | $d_1$ | MLE | Simulated entropy |
|---|---|---|---|---|---|---|
| 1 | 1.0000 | 1.0000 | 0.4997 | 1.0000 | 0.3761 | 1.4742 |
| 2 | 1.0000 | 0.7884 | 0.6435 | 0.6665 | 0.3713 | 1.0709 |
| 3 | 0.8661 | 1.0000 | 0.3934 | 0.9903 | 0.3607 | 1.15806 |
| 4 | 0.7746 | 0.6588 | 0.5846 | 0.4931 | 0.3460 | 1.1133 |
| 5 | 1.0000 | 0.7000 | 0.6780 | 0.1069 | 0.3437 | 0.7281 |
| 6 | 1.0000 | 0.7000 | 0.7000 | 0.2542 | 0.3425 | 1.16843 |
| 7 | 0.7743 | 0.6716 | 0.5892 | 1.8469 | 0.3391 | 1.5712 |
| 8 | 0.9248 | 0.7491 | 0.6686 | 0.6814 | 0.3385 | 1.1628 |
| 9 | 0.7178 | 0.6593 | 0.5546 | 0.2247 | 0.3376 | 0.2925 |
| 10 | 0.7060 | 0.6451 | 0.5523 | 0.2181 | 0.3320 | 0.2765 |
| 11 | 0.7060 | 0.7000 | 0.7000 | 0.7000 | 0.2658 | 1.3312 |

**Table 4.7** Optimized MLE and its associated entropy for generating 5-scrolls

| Case | a | b | c | d | MLE | Entropy simulated | Entropy experi-ment |
|---|---|---|---|---|---|---|---|
| 1 | 1.0000 | 0.7250 | 0.2250 | 1.0000 | 0.6919 | 2.2481 | 2.0131 |
| 2 | 0.9880 | 0.7140 | 0.2050 | 1.0000 | 0.6914 | 2.2962 | 2.1472 |
| 3 | 0.9890 | 0.7300 | 0.2070 | 1.0000 | 0.6908 | 2.2708 | 2.0779 |
| 4 | 0.9910 | 0.6810 | 0.2300 | 0.9810 | 0.6814 | 2.2906 | 2.1175 |
| 5 | 0.9880 | 0.7480 | 0.1890 | 1.0000 | 0.6663 | 1.3800 | 1.9619 |
| 6 | 0.9840 | 0.6810 | 0.2270 | 0.9830 | 0.6651 | 2.3365 | 2.0757 |
| 7 | 0.9890 | 0.6810 | 0.2040 | 0.9790 | 0.6645 | 2.1736 | 2.3032 |
| 8 | 1.0000 | 0.7840 | 0.2000 | 1.0000 | 0.6533 | 2.2628 | 2.3024 |
| 9 | 0.9800 | 0.7960 | 0.1570 | 1.0000 | 0.6523 | 1.3214 | 2.1260 |
| 10 | 1.0000 | 0.7330 | 0.2050 | 1.0000 | 0.6471 | 2.2560 | 2.0287 |
| 11 | 0.7000 | 0.7000 | 0.7000 | 0.7000 | 0.2840 | 2.2352 | 1.9403 |

## 4.7 Generating a 50-Scrolls Chaotic Attractor at 66 MHz

In electronics, the challenge in implementing chaotic oscillators is yet generating as many scrolls as the device capabilities allow it. For instance, experiments realized during the past four years showed the generation of 12-scrolls using commercially available amplifiers and 5-scrolls with an integrated circuit fabricated with technology of 0.5 $\mu$m [43]. The generation of these very few number of scrolls is due to the limitations of the electronic devices, like voltage range and frequency response.

**Table 4.8** Optimized MLE and its associated entropy for generating 10-scrolls

| Case | a | b | c | d | MLE | Entropy simulated | Entropy experi-ment |
|---|---|---|---|---|---|---|---|
| 1 | 1.0000 | 0.5160 | 0.1190 | 1.0000 | 0.8853 | 2.8882 | 2.6302 |
| 2 | 1.0000 | 0.5054 | 0.1140 | 1.0000 | 0.8826 | 2.9032 | 2.6152 |
| 3 | 1.0000 | 0.5130 | 0.1180 | 1.0000 | 0.8792 | 2.8863 | 2.6193 |
| 4 | 1.0000 | 0.5410 | 0.1060 | 1.0000 | 0.8712 | 2.8874 | 2.5166 |
| 5 | 1.0000 | 0.5930 | 0.0840 | 1.0000 | 0.8545 | 2.8664 | 2.4594 |
| 6 | 1.0000 | 0.5160 | 0.1580 | 1.0000 | 0.8438 | 2.9273 | 2.6874 |
| 7 | 1.0000 | 0.6430 | 0.0975 | 1.0000 | 0.8314 | 2.8957 | 2.4891 |
| 8 | 1.0000 | 0.7000 | 0.1160 | 1.0000 | 0.7825 | 2.8788 | 2.6890 |
| 9 | 1.0000 | 0.7995 | 0.2127 | 0.9831 | 0.7249 | 2.6036 | 1.8740 |
| 10 | 1.0000 | 0.7200 | 0.4195 | 1.0000 | 0.6177 | 2.8748 | 2.6213 |
| 11 | 0.7000 | 0.7000 | 0.7000 | 0.7000 | 0.3026 | 2.8956 | 2.6157 |

**Table 4.9** 28-bits computer arithmetic to generate from 10 to 50 scrolls

| Scrolls | Format | Sign | Integer part | Fractional part |
|---|---|---|---|---|
| 10 | 5.23 | 0 | 0000 | .00000000000000000000000 |
| 11 | 5.23 | 0 | 0000 | .00000000000000000000000 |
| 20 | 6.22 | 0 | 00000 | .0000000000000000000000 |
| 30 | 7.21 | 0 | 000000 | .000000000000000000000 |
| 40 | 7.21 | 0 | 000000 | .000000000000000000000 |
| 50 | 7.21 | 0 | 000000 | .000000000000000000000 |

Besides, this section introduces the generation of 50-scrolls using FPGAs and working at 66 MHz.

As mentioned before, the FPGA realization of a dynamical system begins from the description of the discrete equations, where one can identify the required kind of blocks to be described in VHDL, e.g., multipliers, subtractors, comparators, registers, and multiplexers. The VHDL blocks need computer arithmetic notation, which has two fundamental principles: numeric representation and execution of algebraic operations [44]. Using fixed-point notation accelerates the processing speed and saves FPGA resources. The representation is done depending on the number of scrolls to be generated. For example, Table 4.9 shows the numerical format for generating 10, 11, 20, 30, 40, and 50 scrolls, where 28 bits are used as word length.

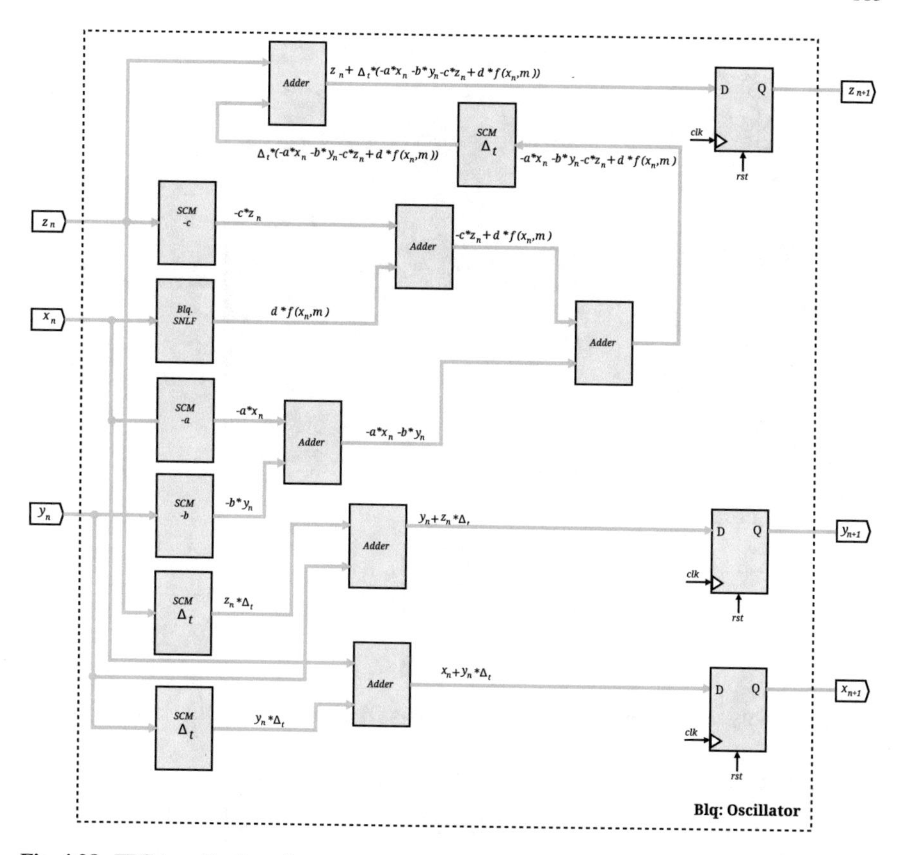


**Fig. 4.28** FPGA realization of $x_{n+1}$, $y_{n+1}$ and $z_{n+1}$

Figure 4.28 shows the implementation of the three state variables, namely: $x_{n+1}$, $y_{n+1}$ and $z_{n+1}$. As one sees, the logic paths for $x_{n+1}$ and $y_{n+1}$ are shorter compared to the one for $z_{n+1}$. In this manner, the output for $x_{n+1}$ and $y_{n+1}$ are computed with lower clock cycles.

To save hardware resources, multipliers are replaced by single constant multiplication (SCM) blocks, which are implemented as already shown in [25]. The state variables are controlled by a counter that generates an activation signal *LDA* every 8 clock cycles, as shown in Fig. 4.29. When $LDA = 0$, the multiplexers load the initial conditions data for the state variables to perform the first iteration to evaluate the equations. Afterwards, $LDA = 1$ and the data in the state variables are then feedback to the multiplexers to continue evaluating until the number of steps is accomplished.

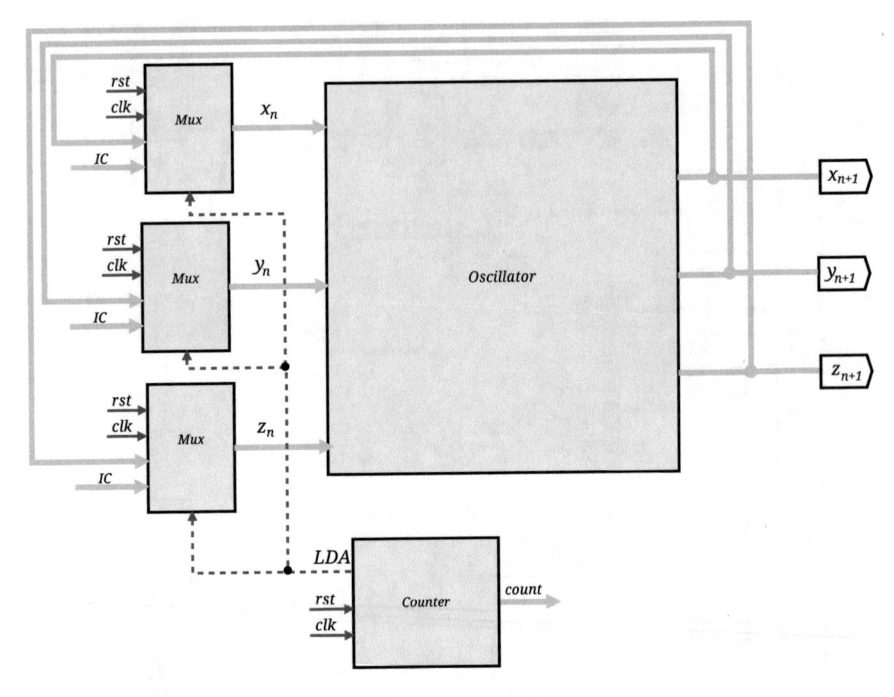


**Fig. 4.29** Implementation of a multi-scroll chaotic oscillator

**Table 4.10** Generating 10-scrolls using the XC3S1000-5FT256 FPGA Spartan-3

| Resources | Available | MLE = 0.16595 | MLE = 0.775137 | MLE = 0.776849 |
|---|---|---|---|---|
| Slice registers | 15,360 | 1,454 | 1,222 | 1,221 |
| Occupied slices | 7,680 | 3,344 | 1,990 | 1,992 |
| 4 input LUTs | 15,360 | 6,129 | 3,483 | 3,487 |
| Bonded IOBs | 173 | 24 | 25 | 25 |
| BUFGMUXs | 8 | 1 | 1 | 1 |

To observe the generation of up to 50-scrolls in an oscilloscope, the XC3S1000-5FT256 FPGA Spartan-3 from Xilinx, and the Cyclone IV GX FPGA DE2i-150 from Altera, were used herein. Table 4.10 lists the used resources for generating 10-scrolls and by applying Forward Euler method for solving the dynamical system.

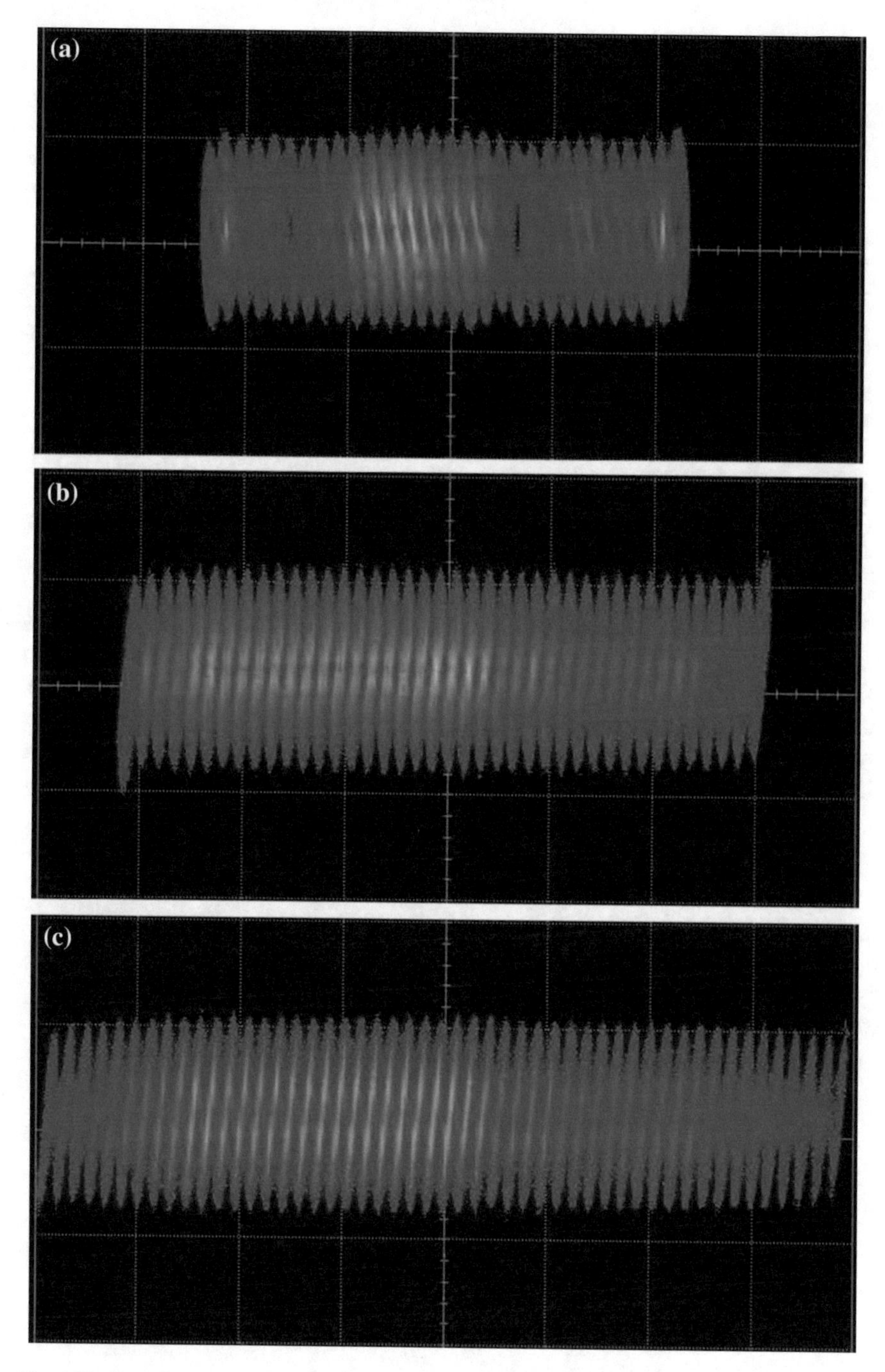


**Fig. 4.30** Experimental results ($x$–$y$ portrait) for generating **a** 30, **b** 40, and **c** 50-scrolls attractors

**Table 4.11** Generating 20, 30, 40, and 50 scrolls using the Cyclone IV GX FPGA DE2i-150

| Scrolls | | 20 | 30 | 40 | 50 |
|---|---|---|---|---|---|
| Resources | Available | Used | Used | Used | Used |
| Dedicated logic registers | 149,760 | 2,222 | 3,237 | 4,281 | 5,235 |
| Total combination function | 149,760 | 9,598 | 15,241 | 20,684 | 25,710 |
| Total logic elements | 149,760 | 9,754 | 15,417 | 20,863 | 25,893 |
| Total pins | 508 | 25 | 25 | 25 | 25 |
| Embedded multiplier 9-bit elements | 720 | 0 | 0 | 0 | 0 |
| Total memory bits | 6,635,520 | 0 | 0 | 0 | 0 |
| Fmax (MHz) | | 78.31 | 66.76 | 66.34 | 66.21 |

Figure 4.30 shows the experimental generation of 30, 40, and 50 scrolls. As the hardware resources increases, in these cases the Cyclone IV GX FPGA DE2i-150 from Altera, is used. Table 4.11 lists the used resources for generating 20, 30, 40, and 50-scrolls. As one sees, the frequency of operation goes down as the number of scrolls increase, so that using the FPGA DE2i-150 from Altera, the frequency of operation for generating 50-scrolls is a little higher than 66 MHz.

# Chapter 5
# Artificial Neural Networks for Time Series Prediction

## 5.1 Introduction

As mentioned in [45], modeling real-world systems plays a pivotal role in their analysis and contributes to a better understanding of their behavior and performance. On the one hand, it has been demonstrated that some diseases of the human body present neurological and behavioral manifestations that can be modeled by chaotic systems. On the other hand, artificial neural networks (ANNs) are powerful tools for modeling and prediction. For instance, the authors in [46] introduced a competitive cooperative coevolution method for training recurrent neural networks for chaotic time-series prediction, [47] describes how to control chaos in a chaotic neural network, [48] proposes the design and analysis of a novel chaotic diagonal recurrent neural network, [49] detects the effect of autapse on coupled neuronal networks, and [50] introduces the design of a feed-forward neuronal network. In this manner, and as recently demonstrated in [51], this chapter shows the hardware implementation of an ANN that is used for chaotic time-series prediction. The cases of study are chaotic time series generated by FPGA-based chaos generators and the prediction is performed for different chaotic signals whose unpredictability is quantified by evaluating their MLE; because the higher the MLE value the higher the unpredictability of an encrypted signal [33, 52].

Reference [53] lists a survey for the fixation of hidden neurons in neural networks for the past 20 years, highlighting the objective to minimize error, improve accuracy and stability. However, no one provided a general solution for implementing an ANN yet, so that the following problems remain: selection of hidden neurons, training algorithm, and architecture. Hidden neurons influence the error on the nodes to which their output is connected, and the accuracy of training is determined by the architecture, number of hidden neurons in hidden layer, kind of activation function, inputs, and updating of weights. These problems are discussed herein taking into account the ANN introduced in [45, 51], and 2 ones designed by applying the geometric pyramid rule.


E. Tlelo-Cuautle et al., *Engineering Applications of FPGAs*,
DOI 10.1007/978-3-319-34115-6_5

Section 5.2 describes the generation of chaotic time series by realizing a chaotic oscillator using an FPGA, and their analysis by evaluating their MLE. Section 5.3 describes the design of an ANN, the activation functions, learning rules, updating of weights, and training algorithms using MATLAB. Three ANN topologies are compared to reproduce chaotic signals with different MLE values. The hardware realization of the ANN is shown in Sect. 5.4, where the activation function is implemented by polynomial and piecewise-linear (PWL) approaches. The FPGA is connected to a personal computer by a serial interface to feed the ANN and experimental results are shown and discussed.

## 5.2 Generating Chaotic Time Series Using FPGAs

A time series represents a measure of a physical variable $x_t$ registered at a time $t$, and it is discrete. The observed data can be used to model or to predict future values to determine the behavior of a time series, as shown in [54], where five sets of chaotic time series data are analyzed. In this chapter, chaotic time series are generated by FPGAs. Again, the chaotic oscillator based on saturated function series is used, which is described by (5.1) [33], where $x$, $y$, and $z$ are the state variables, and $a$, $b$, $c$ and $d_1$, are real and positive coefficients. The PWL function consists of as many saturated levels as the number of scrolls being generated.

$$\begin{aligned} \dot{x} &= y \\ \dot{y} &= z \\ \dot{z} &= -ax - by - cz + d_1 f(x) \end{aligned} \tag{5.1}$$

The pseudocode 1 solves (5.1) by forward Euler, where the discretized state variables are given in (5.2). To generate 2-scrolls, the initial conditions are set to $x_n$, $y_n$, $z_n$ ($x_0 = y_0 = z_0 = 0.01$), the coefficients $a = b = c = d_1 = 0.7$, $\Delta_t = 0.01$ the step size, $m = 60.606$ the slope of the PWL function, $k = \pm 1$ for the saturated levels, and $\alpha = \pm 0.0165$ for the break-points.

For implementing the chaos generator into an FPGA, the numerical representation for the VHDL programming is listed in Table 5.1 as fixed point using 26 bits. The PWL function is included into (5.2), according to (5.3). Figure 5.1 shows the comparator block, where $p01155 = m \cdot d_1 = 0.01155$. Figure 5.2 shows the hardware realization to compute $z_{n+1}$ in (5.2), and Fig. 5.3 to compute $x_{n+1}$ and $y_{n+1}$, where one can appreciate the logic paths. For this dynamical system, a clock signal *clk* is counted eight times to generate an activation signal *LDA* to observe the outputs, mainly $z_{n+1}$. The whole architecture of the chaotic oscillator is shown in Fig. 5.4. The main block *Oscillator* is connected to three multiplexers, whose selection is through signal $P$. Thus, when $P = 0$, the data at IC (initial conditions) for the state variables are loaded. Afterwards, once LDA is activated (after 8 clock cycles), $P = 1$ and then the multiplexers are activated to enable the iterations solving (5.2) [33].

**Table 5.1** Fixed point notation using 26 bits

| Sign | Integer part | Fractional part |
|---|---|---|
| 0 | 000 | 0.1000000000000000000000 |

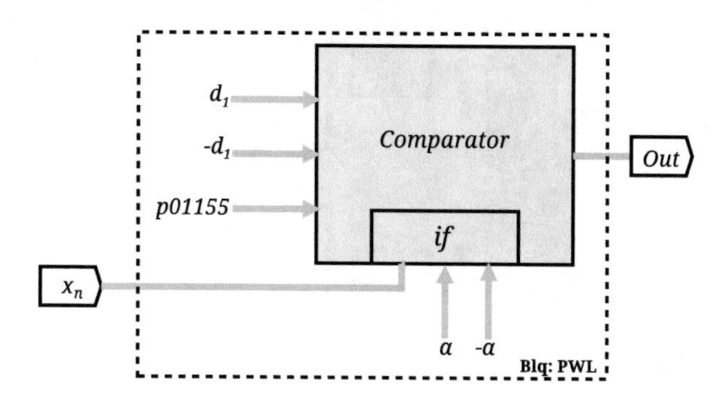


**Fig. 5.1** Block diagram of the comparator

**Pseudocode 1.** Solving (5.1) by applying the Forward Euler numerical method

1: **function**$(x_{n+1}, y_{n+1}, z_{n+1})$= two scrolls $(x_n, y_n, z_n)$
2: $x_{n+1} = x_n + \Delta_t(y_n);$
3: $y_{n+1} = y_n + \Delta_t(z_n);$
4: **if** $x_n > \alpha$
5: $z_{n+1}=z_n + \Delta_t(-ax_n - by_n - cz_n + d_1k);$
6: **elseif** $x_n >= -\alpha \;\&\; x_0 <= \alpha$
7: $z_{n+1}=z_n + \Delta_t(-ax_n - by_n - cz_n + d_1m);$
8: **elseif** $x_n < -\alpha$
9: $z_{n+1}=z_n + \Delta_t(-ax_n - by_n - cz_n - d_1k);$
10: **end**
11: **end**

$$\begin{aligned} x_{n+1} &= x_n + \Delta_t y_n \\ y_{n+1} &= y_n + \Delta_t z_n \\ z_{n+1} &= z_n + \Delta_t(-ax_n - by_n - cz_n + d_1 f(x_n)) \end{aligned} \tag{5.2}$$

$$z_{n+1} = \begin{cases} z + \Delta_t(-ax - by - cz + d_1k) \text{ if } x < \alpha \\ z + \Delta_t(-ax - by - cz + d_1m) \text{ if } -\alpha \leq x \leq \alpha \\ z + \Delta_t(-ax - by - cz - d_1k) \text{ if } x < -\alpha \end{cases} \tag{5.3}$$

To increase the processing speed and to save hardware resources, the multipliers are replaced by single constant multiplication (SCM) blocks [55, 56]. For example, $y = 71x$ when $x = 01_2$ needs the execution of six shifts $x \ll 6 = 01000000_2$, two shifts $x \ll 2 = 0100_2$, and one shift $x \ll 1 = 010_2$. Their sum is $01000110_2 = 70_{10}$, and finally $70_{10} + x = 01000111_2 = 71_{10}$. The multiplications by the constants $a, b, c, d_1 = 0.7$ and $\Delta_t = 0.01$ are then realized by SCMs, as shown in Fig. 5.5.

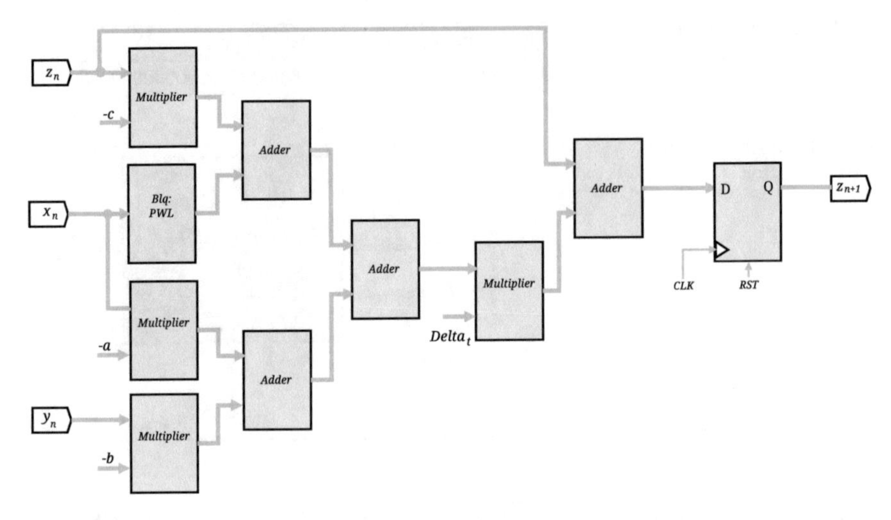


Fig. 5.2 Block description to compute $z_{n+1}$ from (5.2)

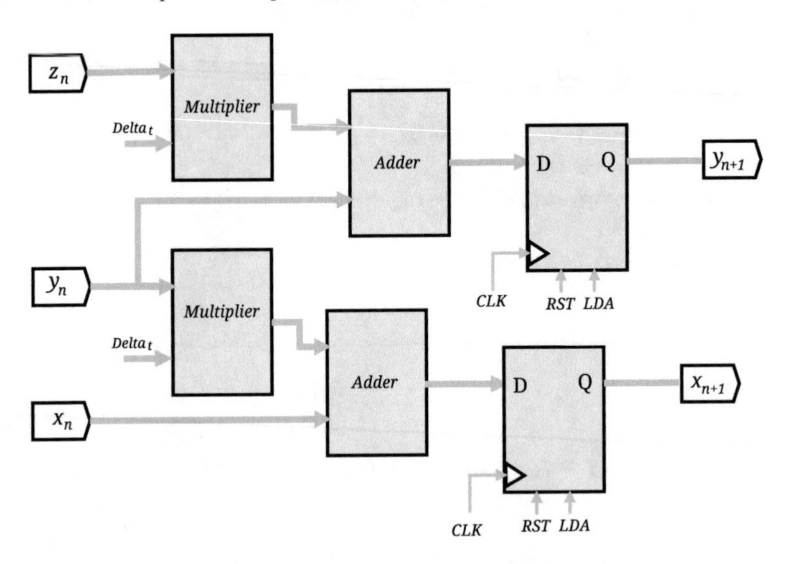


Fig. 5.3 Block description to compute $x_{n+1}$ and $y_{n+1}$

The chaotic oscillator was implemented using Quartus II to perform the FPGA synthesis by VHDL within the Altera Cyclone IV GX FPGA DE2i-150. This FPGA is biased at 12 V, it offers 150 k logic elements, a clock = 50 MHz, 720 M9K blocks, 6480 kbits of on-chip memory, and 360 18 × 18 multiplexers. As shown in the previous chapter, the grade of unpredictability of a chaotic time series is associated to the value of its MLE [33, 52]. That way, Table 5.2 lists the computed MLE for Fig. 5.6. Figures 5.7 and 5.8 show the chaotic time series for MLE = 0.3761 and MLE = 0.3425, respectively.

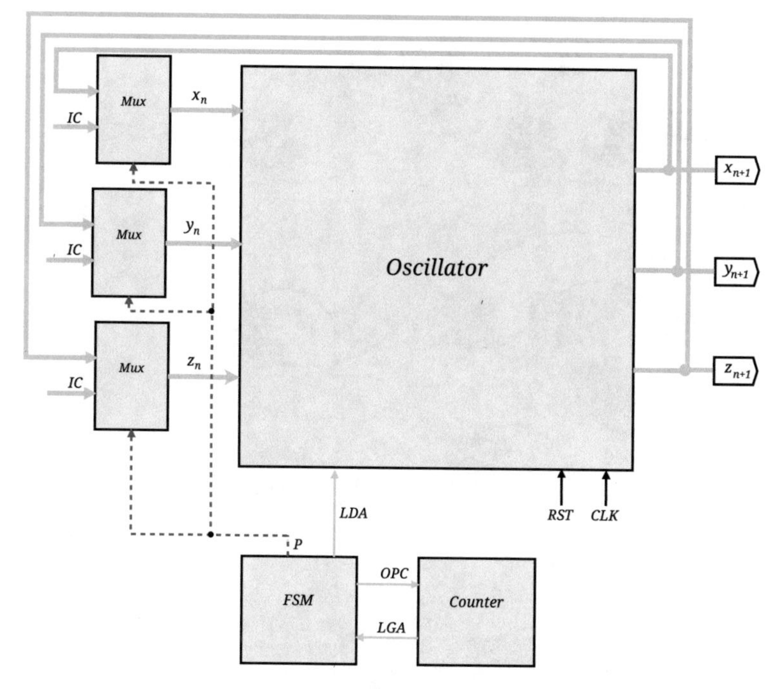


**Fig. 5.4** Block description of the chaotic oscillator

## 5.3 ANN Design Issues

An ANN is a set of elementary processing units called neurons or nodes whose processing capability is stored in the connections by synaptic weights, and whose adaptation depends on learning [57]. There are three kinds of neurons: input (allocate input values), hidden (perform operations and consists of one or more layers), and output ones (perform operations and compare the values with target or reference ones). Figure 5.9 shows the structure where the $j$ neurons receive input signals $x_j$, and $w$ represents the synaptic weights. $b_1$ is the bias and $f(.)$ denotes the activation function that defines the output of the neuron [57]. The state of a neuron $j$ is evaluated by summing the weights, each synaptic weight multiplies each input and since in a soma the inputs coming from the dendrites are added, one gets (5.4). $w_{0j}$ is the excitation threshold, and if it is included at the input the internal state is described by (5.5), with $x_0 = 1$ as a dummy. The internal state is evaluated by (5.6).

$$Yin_j = w_{0j} + \sum_{i=1}^{n} x_i w_{ij} \tag{5.4}$$

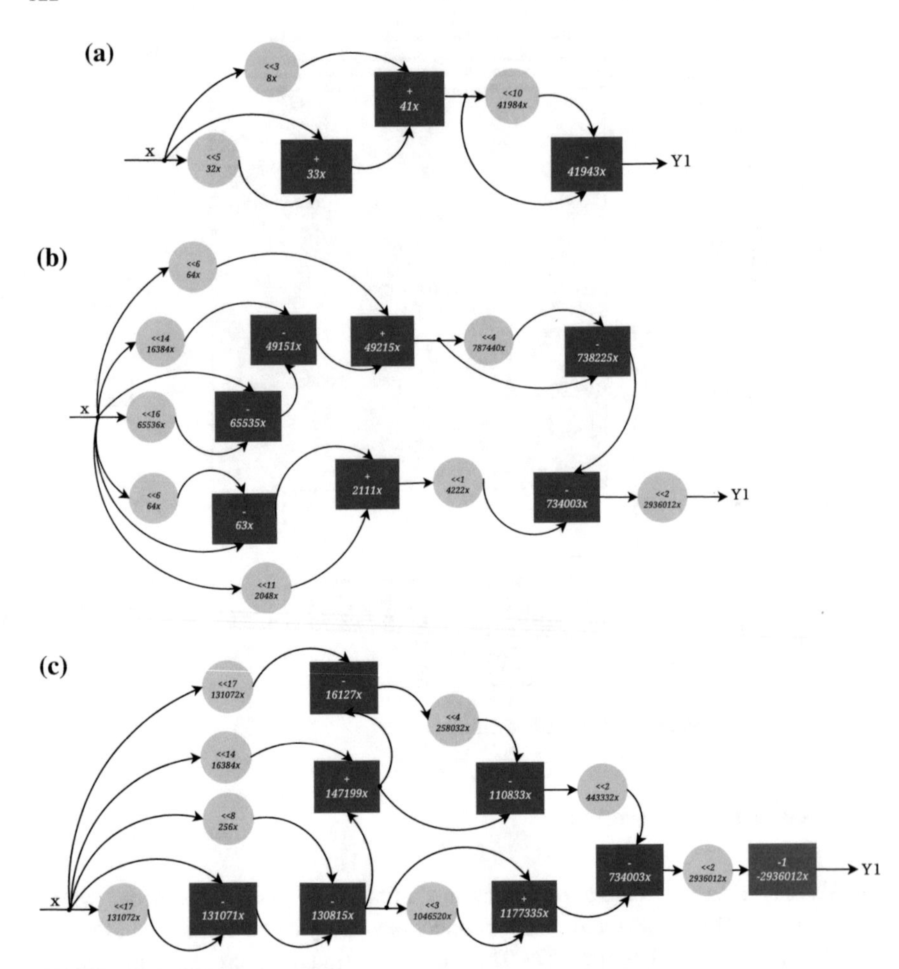


**Fig. 5.5** SCMs multiplying by: **a** $0.01_{10}$, **b** $0.7_{10}$, and **c** $-0.7_{10}$

**Table 5.2** Three different MLE values for the chaotic oscillator described by (5.1)

| Coefficients $a$, $b$, $c$, and $d_1$ | MLE |
|---|---|
| 1.0000, 1.0000, 0.4997, 1.0000 | 0.3761 |
| 1.0000, 0.7000, 0.7000, 0.2542 | 0.3425 |
| 0.7000, 0.7000, 0.7000, 0.7000 | 0.2658 |

$$Yin_j = \sum_{i=1}^{n} x_{ij} w_{ij} \tag{5.5}$$

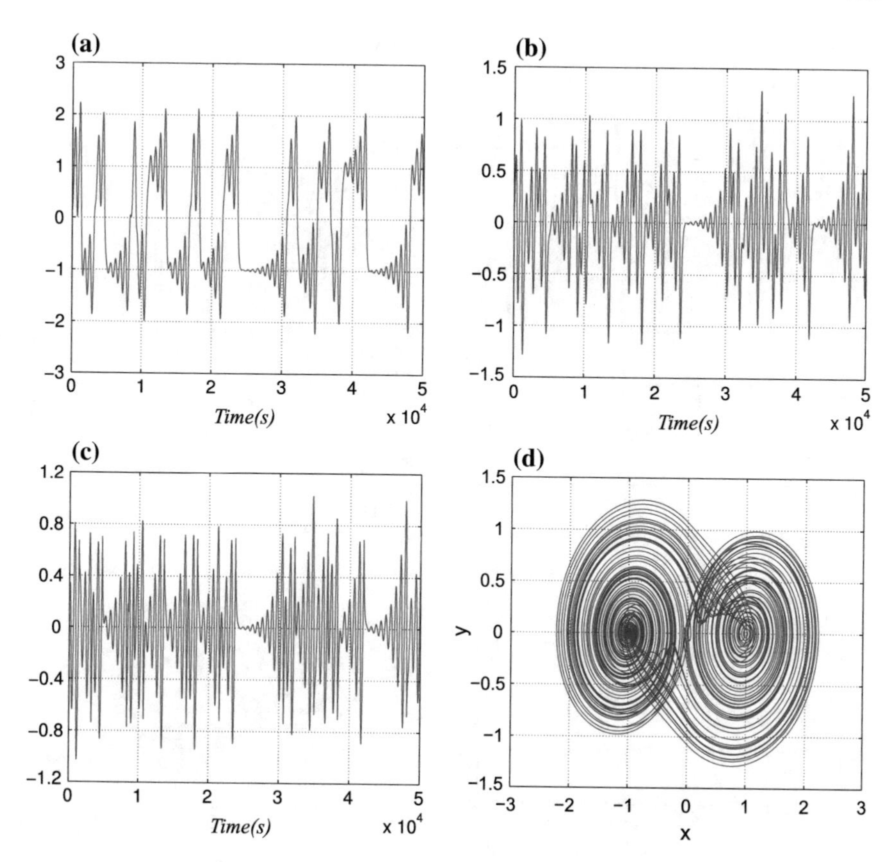


**Fig. 5.6** Chaotic time series with MLE = 0.2658. State variables **a** $x$, **b** $y$, **c** $z$, and **d** phase space portrait $x$–$y$

$$u_j = \sum_{i=1}^{n} x_{ij} w_{ij} + B_j \tag{5.6}$$

If $w_i$ is positive, it is associated to an excitation and if it is negative, to an inhibition. Were the threshold activation incorporated to the weights vector, the output activation $y$ is given by (5.7). $f$ denotes the activation function, and the more used ones are listed in Table 5.3. The neuron needs a learning technique to adjust its parameters during a training process. It can be supervised or unsupervised (self-organized). The supervised one is more suited for discrete data. These and other issues are selected heuristically according to the application.

$$y = f(u) \tag{5.7}$$

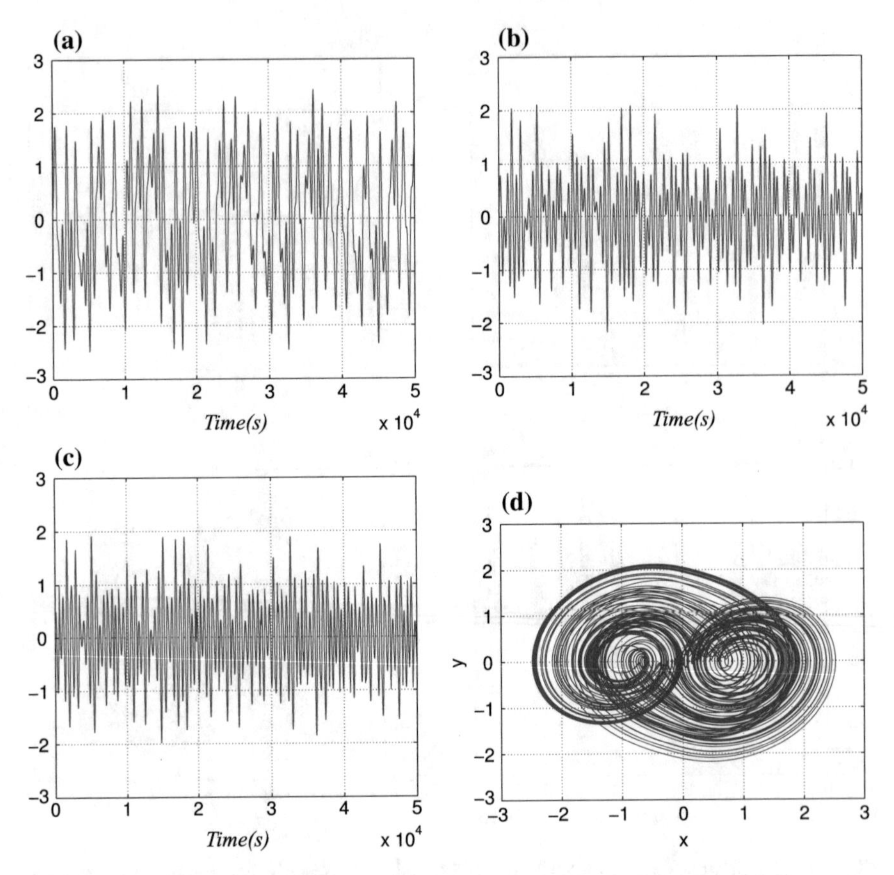


**Fig. 5.7** Chaotic time series with MLE = 0.3761. State variables **a** $x$, **b** $y$, **c** $z$, and **d** phase space portrait $x$–$y$

## 5.3.1 ANN Topology Selection

Up to now, many recipes have been proposed to select the number of hidden layers and neurons in an ANN. For instance, [53] lists a history, but still many researchers apply the geometric pyramid rule [58]. In this chapter, three cases are reviewed to select the ANN topology to be realized into an FPGA.

The first topology consists of three layers: one hidden layer, one output neuron ($m = 1$), and four input neurons ($n = 4$). Then the number of hidden neurons is $h = \sqrt{m \cdot n} = 2$. It also accomplishes the condition that $n > h$. Figure 5.10 shows this topology.

The second topology consists of four layers: two hidden layers, one output neuron ($m = 1$), and eight input neurons ($n = 8$). Now, to determine the number of neurons in the first ($h_1$) and second ($h_2$) hidden layers one should evaluate: $h_1 = m \cdot r^2$ and $h_2 = m \cdot r$, with $r = \sqrt[3]{\frac{n}{m}}$. Figure 5.11 shows the second topology.

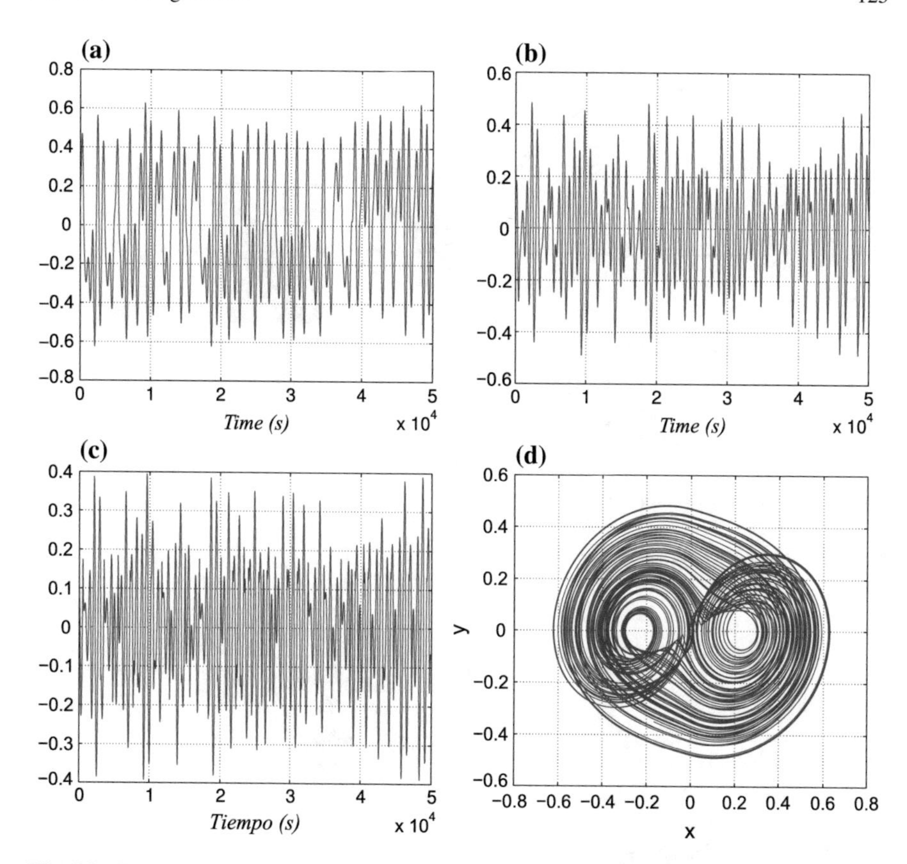


**Fig. 5.8** Chaotic time series with MLE = 0.3425. State variables **a** $x$, **b** $y$, **c** $z$, and **d** phase space portrait $x$–$y$

**Table 5.3** Activation functions

| | |
|---|---|
| Unitary step function | $\phi(u) = \begin{cases} 1 & \text{when } u > 0 \\ 0 & \text{otherwise} \end{cases}$ |
| Lineal function | $\phi(u) = u$ |
| Sigmoid function | $\phi(u) = \dfrac{a}{1 + exp(-bu)}$ |
| Hyperbolic tangent function | $\phi(u) = \dfrac{e^u - e^{-u}}{e^u + e^{-u}}$ |

The third topology is shown in Fig. 5.12, it applies an hyperbolic tangent as activation function, and at the output layer a lineal function. The weights for the input neurons have associated a delay line to provide a finite dynamic response for the time series data.

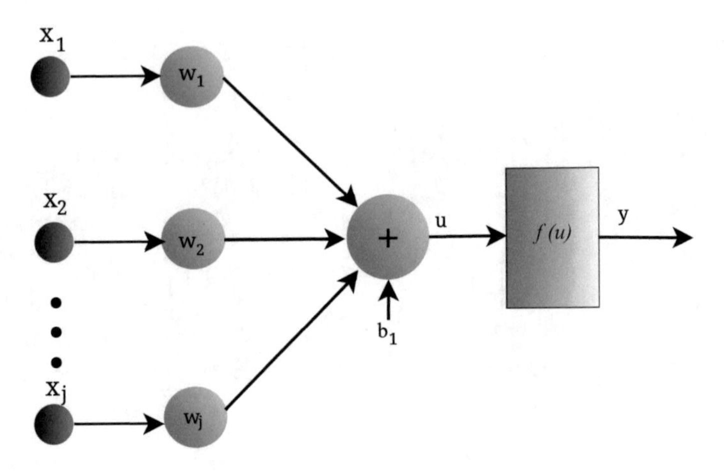


**Fig. 5.9** Functional structure of a neuron

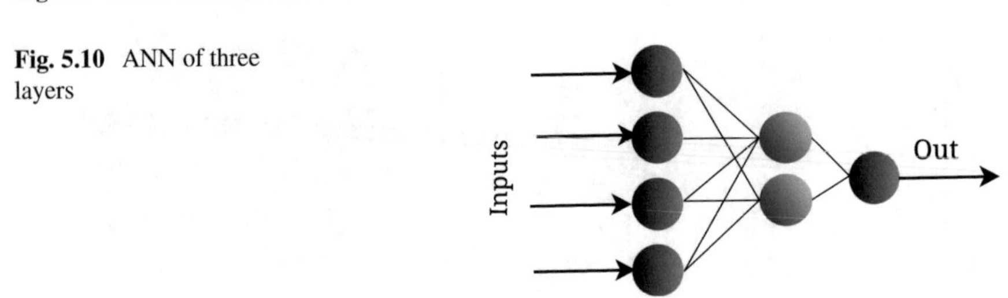


**Fig. 5.10** ANN of three layers

## 5.3.2 *ANN Training*

The batches and the incremental methods are applied to update the weights of the neurons. Among the training algorithms available into MATLAB, the gradient descent with momentum *(traingdm)* is applied herein. The chaotic time series data are generated as shown in Sect. 5.2. When using cheap oscilloscopes the data should be filtered, in our case we applied the MATLAB FIR filter *Savitzky-Golay*, e.g., $y = sgolayfilt(x, k, f)$, with $k = 9$ and $f = 2051$. Figure 5.13 shows the filter response for an experimental chaotic attractor.

Before training, the three ANN topologies shown in Sect. 5.3.1, the chaotic time series data is normalized by MATLAB *mapminmax* within the range $[-1, 1]$. Figure 5.14 shows the normalized values for the experimental data of the state variable $x$ of the chaotic oscillator described by (5.1).

The training is executed using three subsets of data. The first one (training) computes the gradient, weights and bias updating. The second subset (validation) monitors the error during the training. The third subset (test) adjusts the error during the validation process. Such processes are executed into MATLAB *dividerand* using the following values: TRAINING RATIO (TRAINRATIO) $= 0.8$, VALIDATION RATIO (VALRATIO) $= 0.1$, TEST RATIO (TESTRATIO) $= 0.1$.

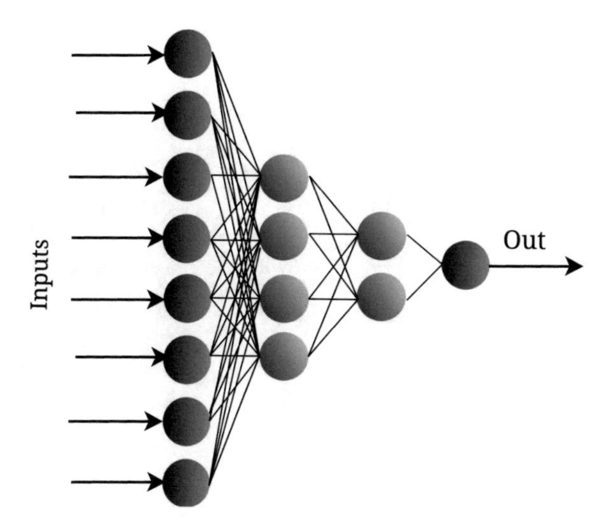


**Fig. 5.11** ANN of four layers

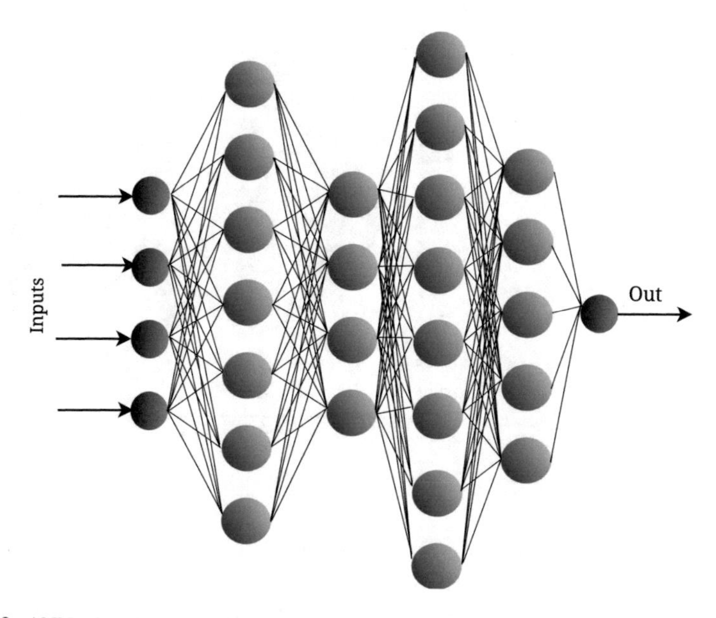


**Fig. 5.12** ANN taken from [45, 51]

Applying the gradient descent with momentum, the training is performed by modifying the values of the learning rule LR and moment constant MC. The search range is [0, 1] with steps of 0.1, and 30 experiments were performed for each possible combination. The ANN performance is evaluated considering the training time, number of epochs for weights updating and the minimum square error (MSE). Table 5.4 lists the ANN training results using experimental data. Figures 5.15, 5.16 and 5.17 show

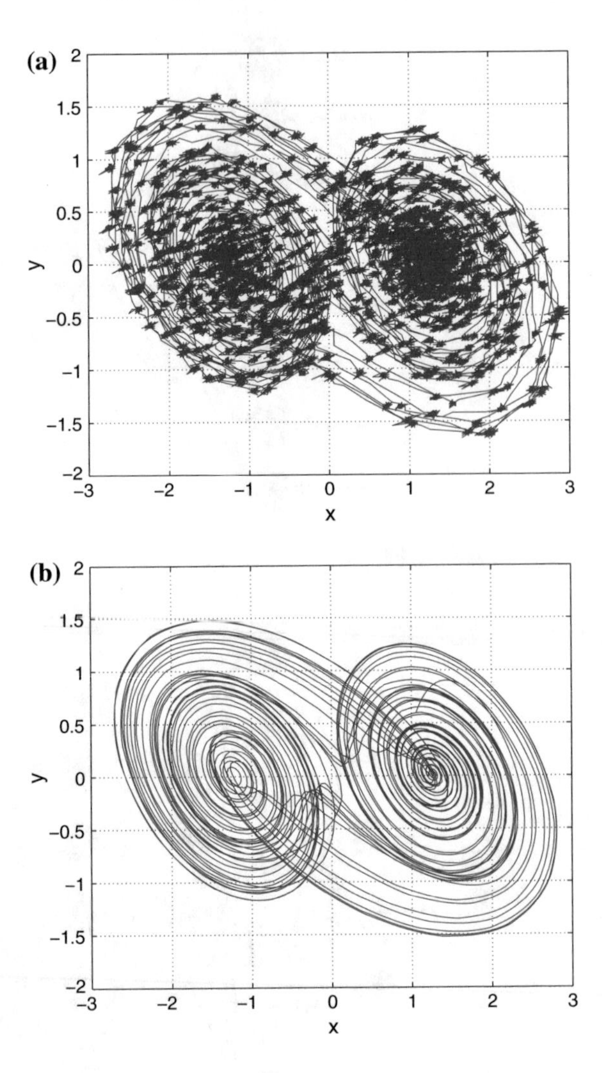


**Fig. 5.13** 2-scrolls attractor experimental data: **a** before and, **b** after filtering

the results to highlight that the ANN with the lower number of epochs, training time, and MSE is the one with six layers, and it will be used in the following experiments:

In the experimental results shown above, a delay line $\Delta = 3$ was used. Now, using the ANN with six layers, Fig. 5.18 shows results with different values for learning rule (LR), moment constant (MC) and $\Delta$, observing that the better performance is when LR $= 0.7$ and MC $= 0.9$. Table 5.5 lists the values for $\Delta$ providing low MSE.

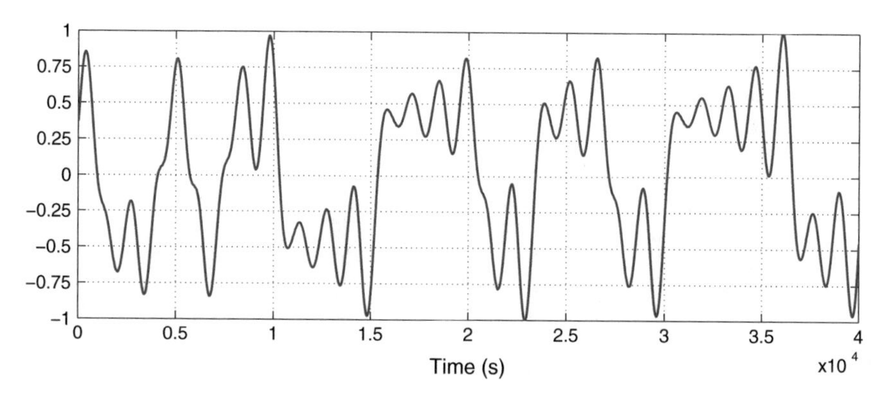


**Fig. 5.14** Experimental data for the chaotic state variable $x$ from (5.1)

**Table 5.4** Average training results for the ANN topologies in Sect. 5.3.1

| | Three layers ANN | | | Four layers ANN | | | Six layers ANN | | |
|---|---|---|---|---|---|---|---|---|---|
| LR, MC | Epoch | Time | MSE | Epoch | Time | MSE | Epoch | Time | MSE |
| 1.0, 0.9 | 765.90 | 154.6087 | 0.000135 | 229.05 | 42.3749 | 0.02626 | 206.70 | 62.0141 | 0.01406 |
| 0.9, 0.8 | 651.55 | 131.6434 | 0.0262000 | 197.35 | 36.7112 | 0.01371 | 262.50 | 80.49775 | 0.001620 |
| 0.7, 0.9 | 854.05 | 173.2616 | 0.013180 | 462.10 | 96.92925 | 0.026185 | 158.50 | 48.5908 | 0.00058 |
| 0.7, 0.8 | 817.95 | 165.9386 | 0.000165 | 251.95 | 46.81955 | 0.02649 | 308.70 | 91.2538 | 0.000985 |
| 0.6, 0.9 | 1405.15 | 321.6969 | 0.000140 | 354.80 | 74.6030 | 0.02623 | 191.30 | 60.1968 | 0.00057 |
| 0.6, 0.7 | 915 | 185.2244 | 0.026260 | 233.10 | 49.1665 | 0.01387 | 285.65 | 88.0419 | 0.01381 |
| 0.5, 0.6 | 818.10 | 202.8650 | 0.065255 | 217.35 | 45.9118 | 0.02747 | 287.05 | 89.79685 | 0.065275 |
| 0.3, 0.9 | 2177.45 | 557.2294 | 0.026250 | 1325.45 | 281.012 | 0.05220 | 299.10 | 95.1104 | 0.01330 |
| 0.2, 0.9 | 2144.45 | 426.3107 | 0.143375 | 1265.75 | 341.2769 | 0.09127 | 330.85 | 99.0527 | 0.10432 |
| 0.2, 0.8 | 1425.15 | 262.0318 | 0.130285 | 1711.95 | 442.9853 | 0.05222 | 361.75 | 107.5476 | 0.013735 |
| 0.1, 0.7 | 2583.40 | 626.1664 | 0.169280 | 2586.3 | 680.9546 | 0.09123 | 484.65 | 141.421 | 0.07881 |

**Table 5.5** Delay lines with better performance using LR $= 0.7$ and MC $= 0.9$

| $\Delta$ | MSE | Time (s) | Epoch |
|---|---|---|---|
| 2 | 179.8 | 44.17 | 0.00199 |
| 3 | 169.7 | 42.13 | 0.00037 |
| 5 | 189 | 46.06 | 0.00146 |
| 7 | 193.1 | 48.39 | 0.0002515 |
| 8 | 194.2 | 47.64 | 0.0006578 |
| 9 | 193.3 | 40.89 | 0.0006578 |

### 5.3.3 *Weights Updating by Batches and Incremental Methods*

MATLAB allows performing the neuron weights updating of both batches and incremental methods. The batch one uses the procedures *adapt* or *train*. The second, *train*

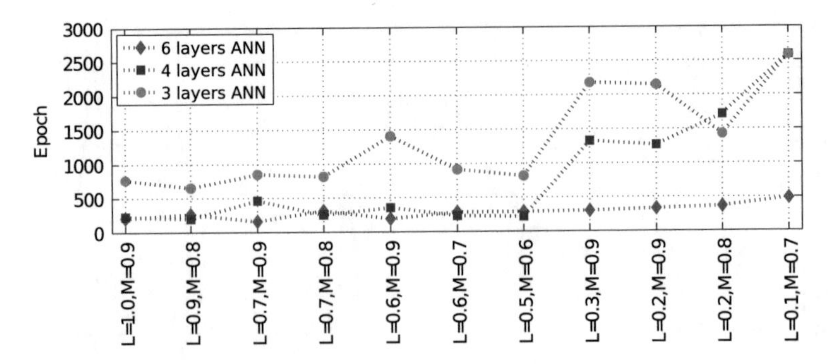


**Fig. 5.15** Average of the number of epochs

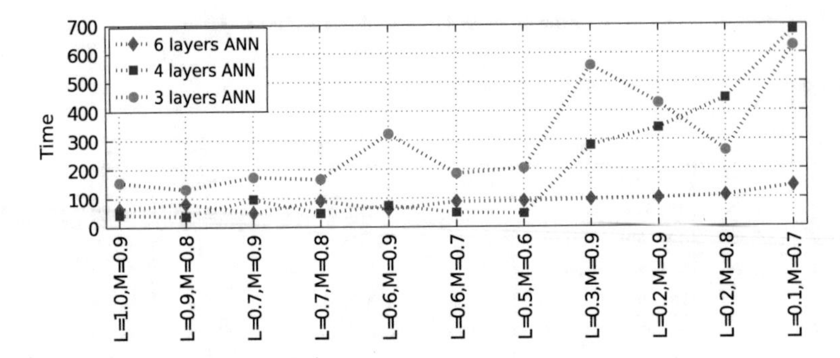


**Fig. 5.16** Average of the training time

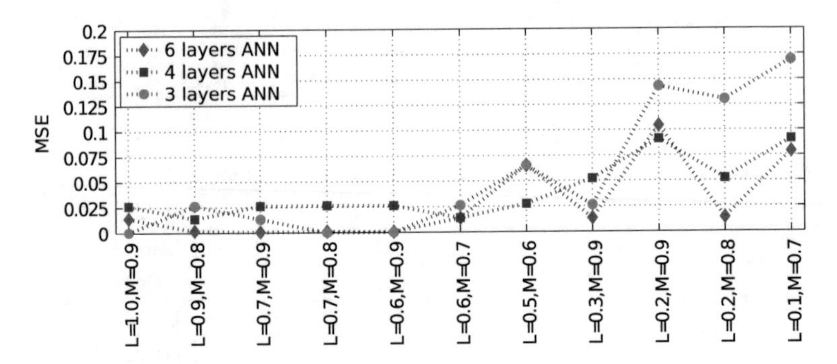


**Fig. 5.17** Average of the MSE value

has access to more training algorithms. The incremental method uses *adapt* depending on the input format that determines the training algorithm, e.g., for an input sequence, the ANN is trained in incremental mode, while for concurrent input vectors the training is by batches.

To select the best weight updating method to perform a better time series prediction, several experiments were executed. In the results using the batches method, the ANN is trained with the same sequence as in the incremental training and the

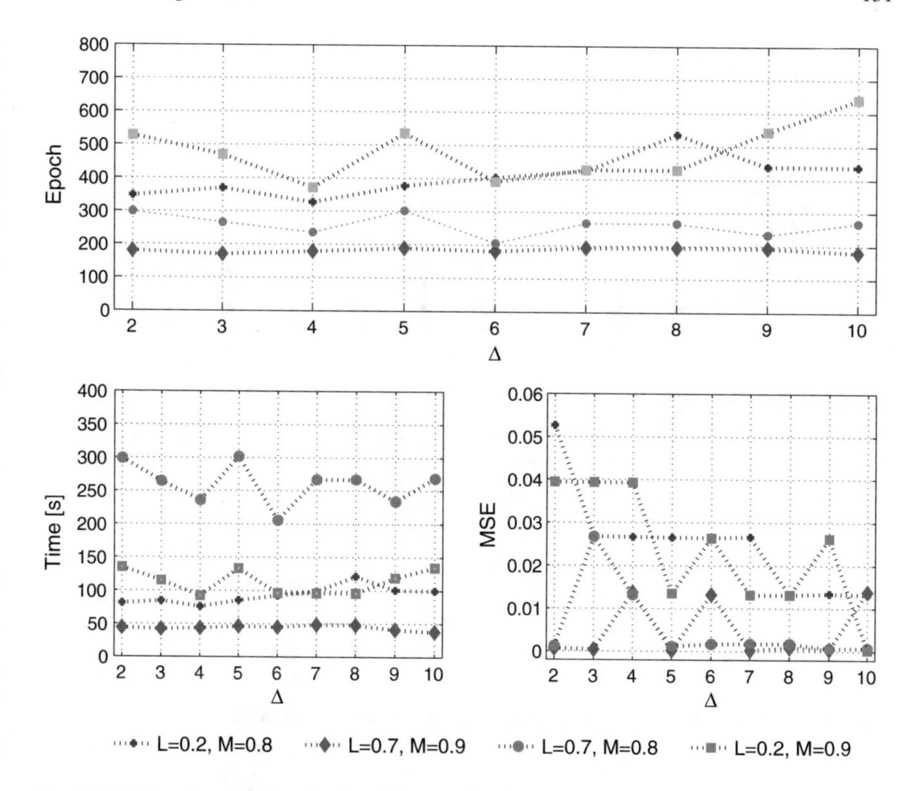


**Fig. 5.18** Delay line variation for the six layers ANN

weights are updated after all inputs are applied (batches), i.e., the ANN is simulated in incremental mode because the input is a sequence, but the weights are updated by batches. Figure 5.19 shows the training time for the ANN with six layers by applying the incremental and batches methods. Figure 5.20 shows the MSE, and one can appreciate that it is low for the training by batches. This is confirmed by predicting a chaotic time series and performing experiments to quantify the training time, as shown by Fig. 5.21.

### 5.3.4 *On the Activation Function in the Last Layer of the ANN*

From the results shown in the previous subsections one can see that the batches method is better than the incremental one when using the ANN with six layers. It reproduces a chaotic time series much better with $\Delta = 3$ with a low MSE. Table 5.6 lists the average values for the gradient, error and epochs for different LR and MC

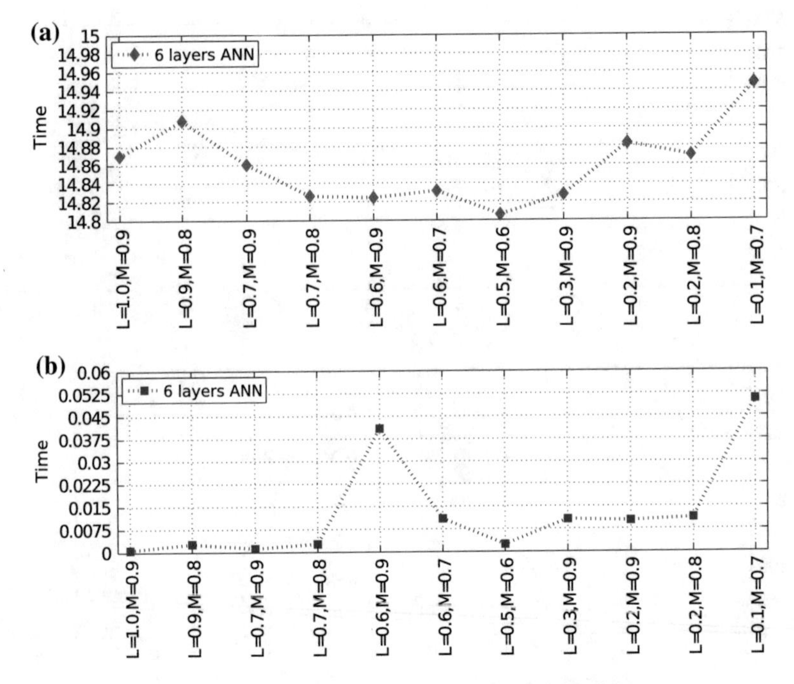


**Fig. 5.19** Training time for weight updating: **a** incremental and, **b** batches

values when applying the gradient descent with momentum algorithm. The activation function is the hyperbolic tangent one for all neurons and h = 0.1.

Using a linear activation function in the neuron in the last layer helps to reduce the number of epochs as shown in Fig. 5.22, where the experiments were realized with five MC values and varying LR with steps ($h$) of 0.1. This is also appreciated in Table 5.7, where it is quite clear that the number of epochs is lower with respect to Table 5.6, just by changing the activation function in the last layer. This difference is better appreciated in Fig. 5.23 with different values for LR and MC.

### 5.3.5 *Time Series Prediction of Chaotic Signals with Different MLE*

The previous results were computed using a chaotic signal with a low value of its maximum Lyapunov exponent (MLE) that is associated to $a = b = c = d_1 = 0.7$ in (5.1) and Table 5.2, where three MLE values are listed. Table 5.8 lists the averages for the three ANN topologies shown in Sect. 5.3.1 when MLE = 0.3761.

Figure 5.24 shows the results for 20 experiments when training the six layers ANN with three chaotic signals and with different MLE value. As one can infer, the MLE

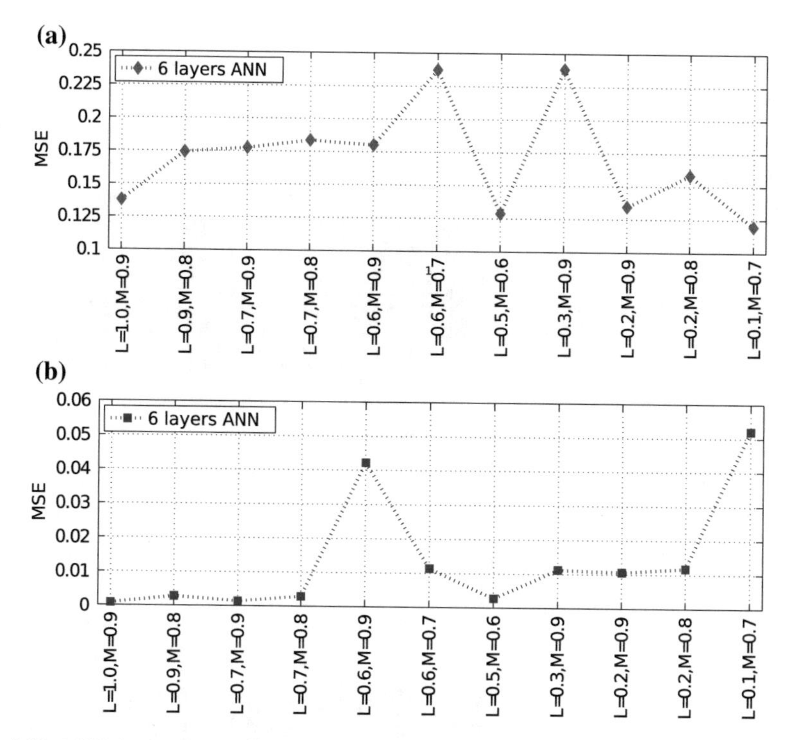


**Fig. 5.20** MSE for weight updating: **a** incremental and **b** batches

is correlated to the training time. In the majority of cases, also the MSE value is correlated to the MLE value, as shown in Fig. 5.25.

## 5.4 FPGA-Based ANN for Time-Series Prediction of Chaotic Signals

The six layers ANN showed better results for predicting chaotic time series with different MLE value. In this manner, this section shows its FPGA realization using the Altera Cyclone IV GX FPGA, DE2i-150.

### *5.4.1 FPGA Realization of the Hyperbolic Tangent Activation Function*

The hyperbolic tangent activation function has a sigmoid behavior that can be approached by a polynomial function. Table 5.9 lists the coefficient values of a poly-

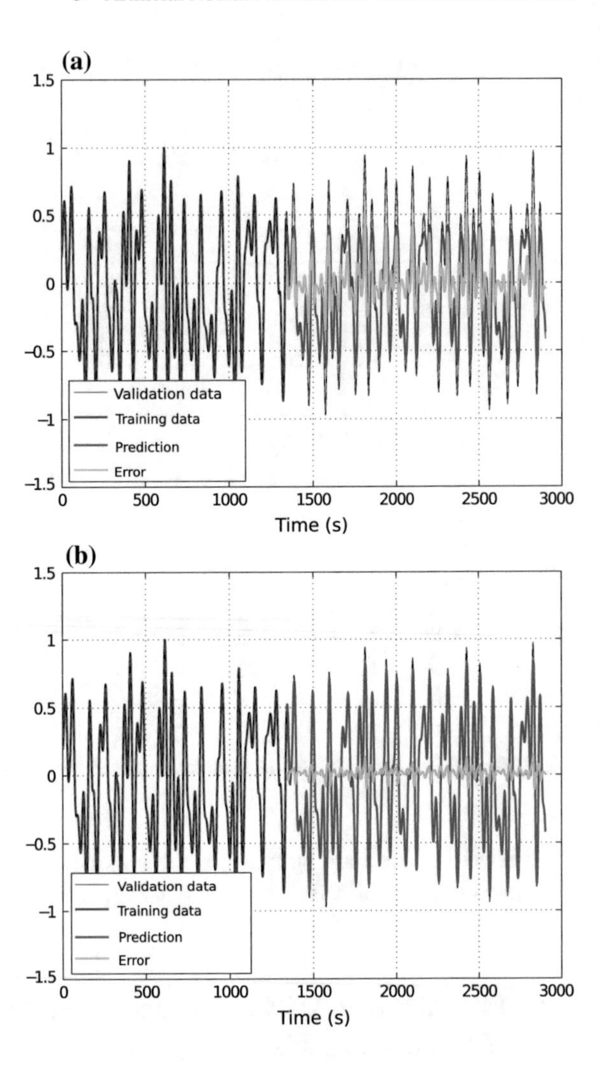


**Fig. 5.21** Time series prediction for the state variable $x$ with LR $= 0.6$, MC $= 0.9$, and $a = b = c = d_1 = 0.7$ in (5.1): **a** incremental and **b** batches

nomial of order 8, 9, and 10. As one sees in Fig. 5.26, they cannot completely approach this activation function because an error arises in the middle or at the extremes of the function. In addition, looking at the coefficient values in Table 5.9, the hardware is expensive when using a lookup table. A better approximation is performed using the PWL approach introduced in [59], which is approached by segments of order 2, as shown in (5.8), where $\beta$ and $\theta$ determine the slope and gain $H_{s1}(z)$ between $-L \leq z \leq L$. In this manner, the hyperbolic tangent is approached herein by (5.9), where $\theta = 0.25$, $L = 2$ and $\beta = 1$. Figure 5.27 shows the comparison when the hyperbolic tangent activation function is approached by polynomials or PWL functions. The superiority of the PWL approach is highlighted and this activation function is realized as shown by Fig. 5.28.

**Table 5.6** Average results when the activation function is hyperbolic tangent in all neurons and h = 0.1

| LR and MC | Gradient average | Error average | Epochs average | LR and MC | Gradient average | Error average | Epochs average |
|---|---|---|---|---|---|---|---|
| 0.9, 0.8 | 0.002497 | 0.5 | 713.1 | 0.4, 0.9 | 0.008425 | 0.9 | 3895 |
| 0.9, 0.7 | 0.000916 | 0.4 | 662.1 | 0.4, 0.8 | 0.010550 | 0.3 | 1335.7 |
| 0.9, 0.6 | 0.004866 | 0.2 | 1183.1 | 0.4, 0.7 | 0.00221 | 0.3 | 735.3 |
| 0.8, 0.9 | 0.059100 | 0.3 | 646.4 | 0.4, 0.6 | 0.002332 | 0.2 | 695.7 |
| 0.8, 0.7 | 0.001235 | 0.4 | 623.6 | 0.3, 0.9 | 0.007762 | 0.9 | 7423.6 |
| 0.8, 0.6 | 0.0303799 | 0.1 | 673.9 | 0.3, 0.8 | 0.0106363 | 0.3 | 2268 |
| 0.7, 0.8 | 0.0085 | 0.5 | 8117 | 0.3, 0.7 | 0.0050 | 0.3 | 1483.7 |
| 0.7, 0.9 | 0.0060 | 0.3 | 1219 | 0.3, 0.6 | 0.0020 | 0.3 | 967 |
| 0.7, 0.6 | 0.0410 | 0.2 | 453.8 | 0.2, 0.9 | 0.0187 | 0.7 | 5363.2 |
| 0.6, 0.9 | 0.0085 | 0.7 | 1108 | 0.2, 0.8 | 0.0059 | 0.4 | 3178.9 |
| 0.6, 0.8 | 0.0072 | 0.4 | 409 | 0.2, 0.7 | 0.0051 | 0.4 | 2357.7 |
| 0.6, 0.7 | 0.0131 | 0.1 | 559.6 | 0.2, 0.6 | 0.0025 | 0.3 | 677.7 |
| 0.5, 0.9 | 0.0235 | 0.7 | 2713 | 0.1, 0.9 | 0.0293 | 0.8 | 12053 |
| 0.5, 0.8 | 0.0038 | 0.5 | 1283 | 0.1, 0.8 | 0.0163 | 0.7 | 9156.7 |
| 0.5, 0.7 | 0.0010 | 0.2 | 740.1 | 0.1, 0.7 | 0.000336 | 0.7 | 9636.8 |

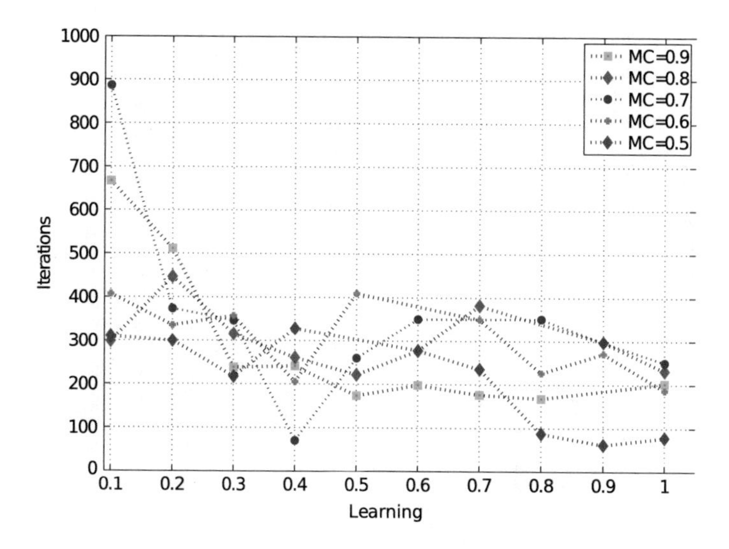


**Fig. 5.22** Iterations with a linear activation function in the last layer

$$H_{s1}(z) = \begin{cases} z(\beta - \theta z) & \text{for } 0 \leq z \leq L \\ z(\beta + \theta z) & -L \leq z \leq 0 \end{cases} \tag{5.8}$$

**Table 5.7** Average results when the activation function is hyperbolic tangent in all neurons except in the last layer, where the activation function is linear

| LR and MC | Gradient average | Error average | Epochs average | LR and MC | Gradient average | Error average | Epochs average |
|---|---|---|---|---|---|---|---|
| 0.1, 0.9 | 0.01522 | 0.7 | 667.09 | 0.5, 0.7 | 0.00350 | 0.6 | 259.9 |
| 0.1, 0.7 | 0.00449 | 0.8 | 886.36 | 0.5, 0.6 | 0.00150 | 0.8 | 408.8 |
| 0.1, 0.6 | 0.00562 | 0.6 | 407 | 0.6, 0.9 | 0.01100 | 0.8 | 198.4 |
| 0.2, 0.9 | 0.01929 | 0.8 | 511.18 | 0.6, 0.8 | 0.00332 | 0.6 | 277.2 |
| 0.2, 0.8 | 0.00408 | 0.9 | 446.6 | 0.6, 0.7 | 0.001090 | 0.8 | 350.4 |
| 0.2, 0.7 | 0.00461 | 0.7 | 373.44 | 0.7, 0.9 | 0.003008 | 0.9 | 176.5 |
| 0.3, 0.9 | 0.001795 | 0.8 | 238.3 | 0.7, 0.8 | 0.001421 | 0.9 | 381.7 |
| 0.3, 0.8 | 0.00297 | 0.5 | 315.1 | 0.7, 0.6 | 0.0063 | 0.4 | 349.7 |
| 0.3, 0.7 | 0.00711 | 0.5 | 346.3 | 0.8,0.9 | 0.0036493 | 0.9 | 168 |
| 0.3, 0.6 | 0.00288 | 0.6 | 357.1 | 0.8, 0.7 | 0.0019718 | 0.4 | 351 |
| 0.4, 0.9 | 0.02190 | 0.6 | 241.11 | 0.8, 0.6 | 2.70e+13 | 0.6 | 226.5 |
| 0.4, 0.8 | 0.01475 | 0.5 | 260.66 | 0.9, 0.7 | 0.0023375 | 0.4 | 297.4 |
| 0.4, 0.6 | 0.00630 | 0.5 | 205 | 0.9, 0.8 | 0.0144457 | 0.3 | 153.3 |
| 0.5, 0.9 | 0.02936 | 0.7 | 173.9 | 1, 0.9 | 0.0051539 | 0.9 | 202.6 |
| 0.5, 0.8 | 0.02967 | 0.5 | 221.7 | 1, 0.6 | 0.0082798 | 0.6 | 250.6 |

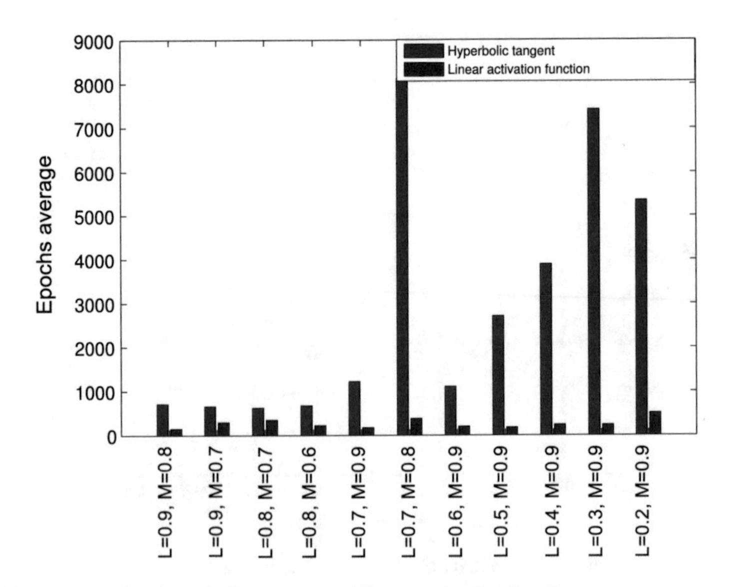


**Fig. 5.23** Epochs using hyperbolic tangent and linear activation functions

**Table 5.8** Average results for three topologies and MLE = 0.3761 from Table 5.2

| | Three layers ANN | | | Four layers ANN | | | Six layers ANN | | |
|---|---|---|---|---|---|---|---|---|---|
| LR, MC | Epoch | Time | MSE | Epoch | Time | MSE | Epoch | Time | MSE |
| 1, 0.9 | 730.80 | 175.411 | 0.00965 | 329.30 | 103.2976 | 0.009559 | 276.65 | 116.405 | 0.000365 |
| 0.9, 0.8 | 988.10 | 255.544 | 0.0001 | 417.85 | 126.0837 | 0.02849 | 230.80 | 95.9772 | 0.00046 |
| 0.7, 0.9 | 1049.8 | 261.082 | 0.019135 | 293 | 91.50 | 0.02875 | 252.20 | 105.040 | 0.00990 |
| 0.7, 0.8 | 1337.40 | 344.44 | 0.009564 | 486.85 | 147.222 | 0.04753 | 269.6 | 116.336 | 0.000365 |
| 0.6, 0.9 | 883.30 | 227.606 | 0.019075 | 287.05 | 86.7168 | 0.000685 | 388.50 | 162.36205 | 0.019195 |
| 0.6, 0.7 | 957.95 | 247.0831 | 0.009765 | 279.05 | 83.9329 | 0.028995 | 300.05 | 127.079 | 0.009785 |
| 0.5, 0.6 | 2082.65 | 586.613 | 0.04763 | 1738.7 | 528.907 | 0.028495 | 392.65 | 169.42 | 0.00971 |
| 0.3, 0.9 | 2729.85 | 964.27 | 0.05705 | 1469.7 | 537.826 | 0.056915 | 725.35 | 311.047 | 0.019105 |
| 0.2, 0.9 | 3247.80 | 899.055 | 0.0665956 | 1328.5 | 486.941 | 0.06651 | 446.6 | 193.0998 | 0.047875 |
| 0.1, 0.7 | 3788.65 | 1379.62 | 0.12350 | 3933.75 | 1466.625 | 0.07588 | 654.650 | 271.989 | 0.07618 |

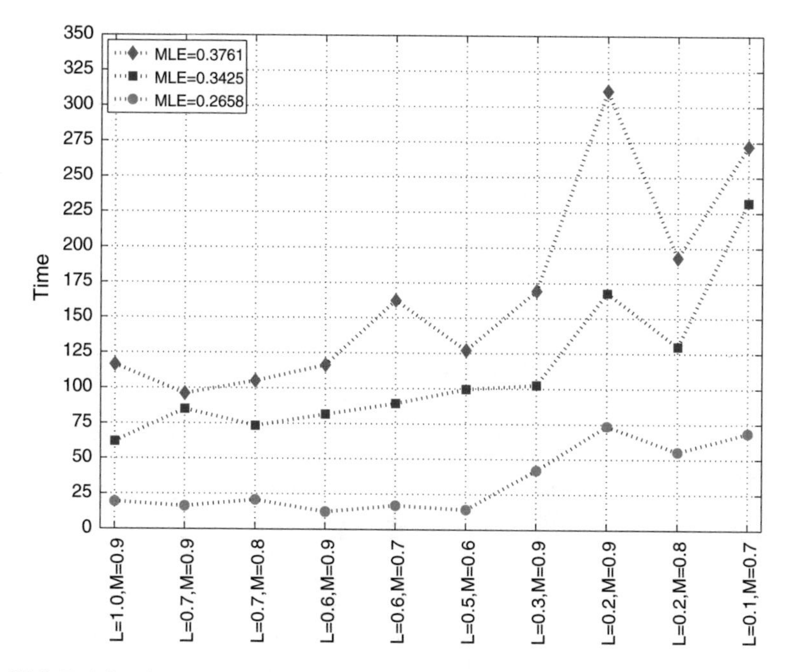


**Fig. 5.24** Training time required for chaotic signals of different MLE value

$$G_{s1}(z) = \begin{cases} 1 & \text{for } L \le z \\ H_{s1} & \text{for } -L < z < L \\ -1 & z \le -L \end{cases} \tag{5.9}$$

The FPGA realization of the whole ANN with six layers is described in the following. The weights are stored in a matrix of the form,

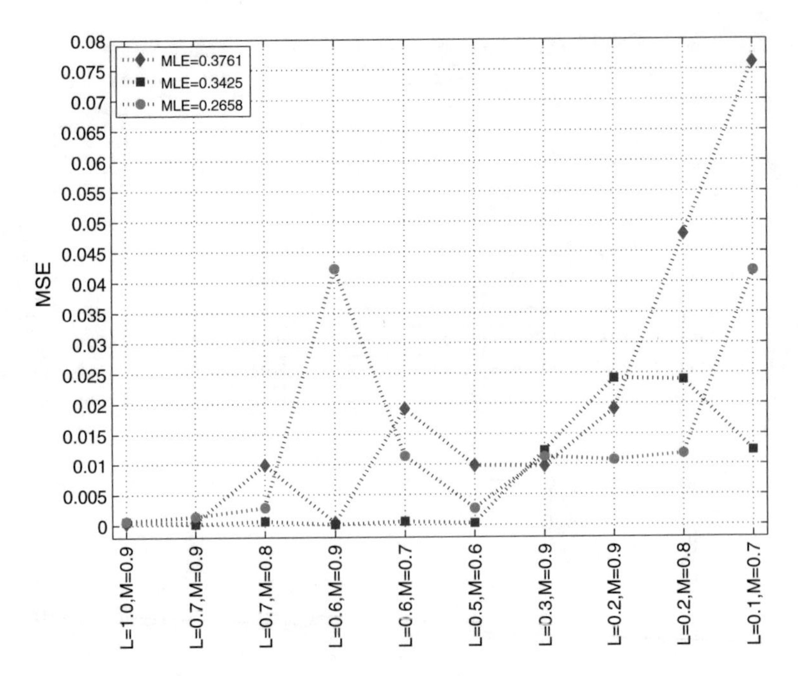


**Fig. 5.25** MSE of the prediction of chaotic signals with different MLE value

**Table 5.9** Polynomial coefficient of order $n$: $ax^n + bx^{n-1}, \ldots, cx^0$

| Order | Polynomial coefficients |
|---|---|
| 8 | 0, −0.00252285599708557, 0, 0.0388774573802948, 0, −0.236943364143372, 0, 0.968010038137436, 0 |
| 9 | 0.000489681959152222, 0, −0.00907894968986511, 0, 0.0678912699222565, 0, −0.283973813056946, 0, 0.988280504941940, 0 |
| 10 | 0, 0.000489681959152222, 0, −0.00907894968986511, 0, 0.0678912699222565, 0, −0.283973813056946, 0, 0.988280504941940, 0 |

$$W = \begin{bmatrix} w_{1,1} & w_{1,2} & \ldots & w_{1,m} \\ w_{2,1} & w_{2,2} & \ldots & w_{2,m} \\ \\ w_{n,1} & w_{n,2} & \ldots & w_{n,m} \end{bmatrix}$$

where $m$ is the element of the input vector and $n$ the neuron of the layer. The biases are allocated in a same way. Table 5.10 lists the weights $W$ and biases $B$ associated to the first layer of the ANN shown in Fig. 5.12. The output of each neuron is given by

$$L_{1,n} = f(In_1 W_{n,1} + In_2 W_{n,2} + In_3 W_{n,3} + B_n) \tag{5.10}$$

where $f()$ corresponds to the activation function of the neurons. The hardware realization is shown in Fig. 5.29, where one can appreciate that the multiplications are

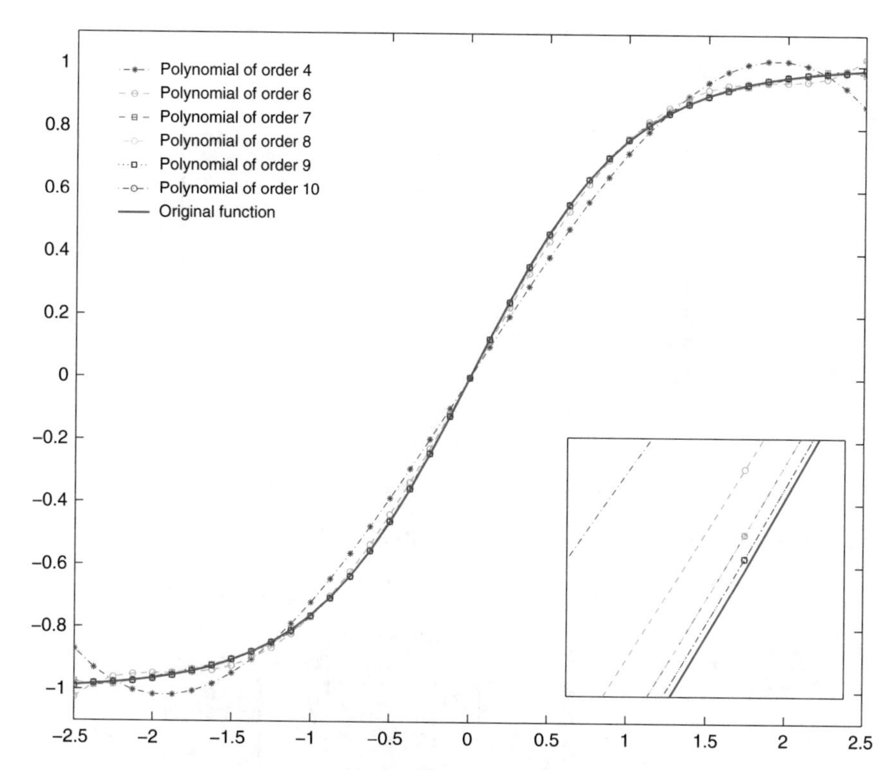


**Fig. 5.26** Hyperbolic tangent function approached by polynomials

**Table 5.10** Weights and biases of the first layer

| | Weights | | | Biases |
|---|---|---|---|---|
| | $m = 1$ | $m = 2$ | $m = 3$ | |
| $n = 1$ | −0.1020998955 | −0.0973799229 | −0.0926499367 | 0.0289199352 |
| $n = 2$ | 0.2051999569 | 0.1905999184 | 0.1758999825 | −0.0622000694 |
| $n = 3$ | 0.0263199806 | 0.0239100456 | 0.0215098858 | 0.0624599457 |
| $n = 4$ | −0.1422998905 | −0.1308000088 | −0.1194000244 | 0.0477199554 |

performed by SCMs to increase the processing speed, as described in Sect. 5.2. The biases were stored in variables with computer arithmetic of 4.22, as shown in Table 5.1.

Table 5.11 shows the weights $W$ and biases $B$ for the second layer in Fig. 5.12. The output of each neuron is given by (5.11), and Fig. 5.30 shows its hardware realization, again the multipliers are implemented by SCMs.

$$L_{2,n} = f(In_1 W_{n,1} + In_2 W_{n,2} + In_3 W_{n,3} + In_4 W_{n,4} + B_n) \tag{5.11}$$

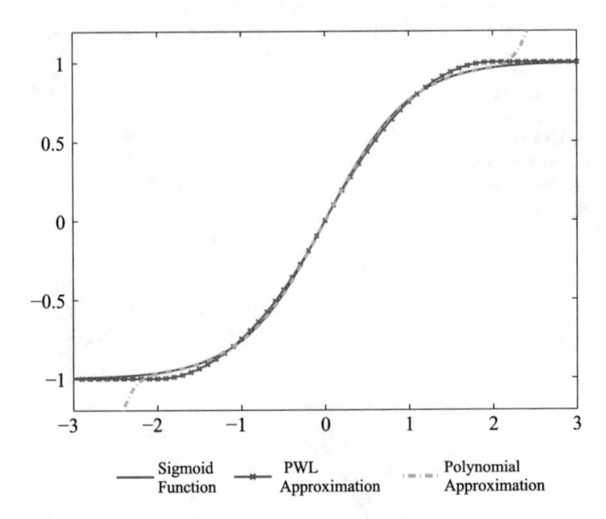


**Fig. 5.27** Approaching the sigmoid behavior by polynomials and PWL functions

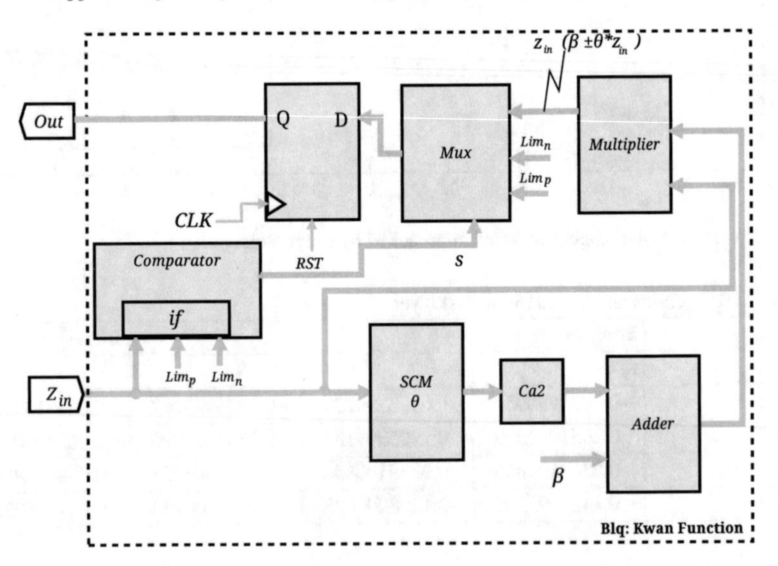


**Fig. 5.28** Hardware realization of the sigmoid from its PWL approach [59]

For the third layer, its output is given by the following equation, and Fig. 5.31 shows the hardware implementation.

$$L_{3,n} = f(In_1 W_{n,1} + In_2 W_{n,2} + In_3 W_{n,3} + In_4 W_{n,4} + In_5 W_{n,5} + In_6 W_{n,6} + In_7 W_{n,7} + B_n) \quad (5.12)$$

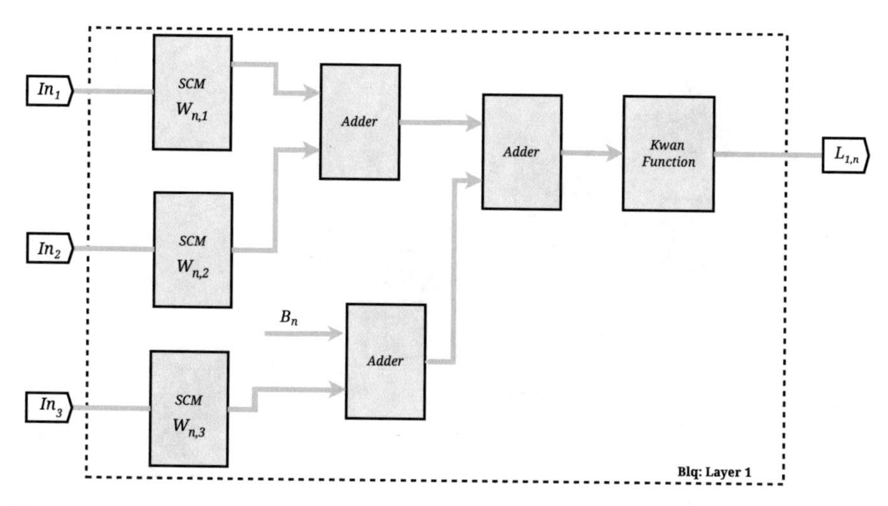


**Fig. 5.29** Hardware of the first layer in Fig. 5.12

**Table 5.11** Weights and biases of the second layer

| | Weights | | | | Biases |
|---|---|---|---|---|---|
| | $m = 1$ | $m = 2$ | $m = 3$ | $m = 4$ | |
| $n = 1$ | −1.6719999313 | −0.4147000313 | 1.4600000381 | 0.2945001125 | 2.2769999504 |
| $n = 2$ | 0.3118999004 | −1.2760000229 | −0.0129299164 | −1.8589999676 | −1.5179998875 |
| $n = 3$ | 0.44810009 | −1.7809998989 | −1.2179999352 | −0.6514000893 | −0.7614998817 |
| $n = 4$ | −0.2053999901 | 0.203099966 | −1.7460000515 | 1.4360001087 | 0.0007128716 |
| $n = 5$ | 0.55189991 | −1.114000082 | 1.7479999065 | 0.8046998978 | 0.7864000797 |
| $n = 6$ | −1.3550000191 | 0.8132998943 | −0.7953000069 | 1.4330000877 | −1.5250000954 |
| $n = 7$ | 0.5044000149 | −0.3478999138 | 2.1930000782 | 0.0106298923 | 2.2769999504 |

The output of the fourth layer is given by the following equation, and its implementation in Fig. 5.32.

$$L_{4,n} = f(In_1 W_{n,1} + In_2 W_{n,2} + In_3 W_{n,3} + In_4 W_{n,4} + B_n) \tag{5.13}$$

The fifth layer has an output given by the following equation, and its implementation by Fig. 5.33.

$$\begin{aligned} L_{5,n} = & f(In_1 W_{n,1} + In_2 W_{n,2} + In_3 W_{n,3} + In_4 W_{n,4} + \\ & In_5 W_{n,5} + In_6 W_{n,6} + In_7 W_{n,7} + In_8 W_{n,8} + B_n) \end{aligned} \tag{5.14}$$

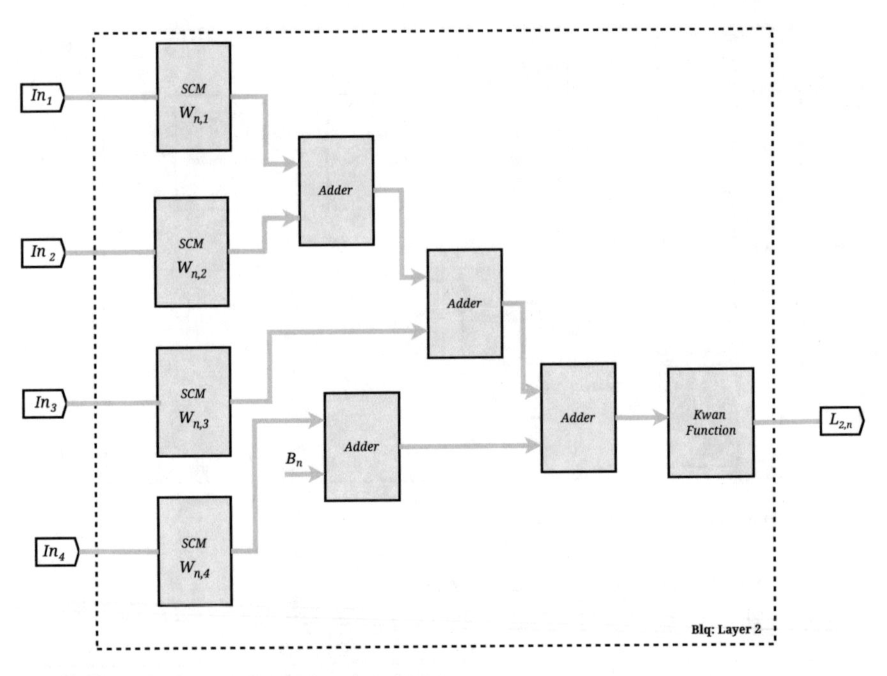


**Fig. 5.30** Hardware of the second layer in Fig. 5.12

**Table 5.12** Weights and biases of the sixth layer

| | Weights | | | | | Biases |
|---|---|---|---|---|---|---|
| | $m = 1$ | $m = 2$ | $m = 3$ | $m = 4$ | $m = 5$ | |
| $n = 1$ | −0.0549499989 | −0.0297698975 | −1.2409999371 | −0.299200058 | 0.0297698975 | 0.0297698975 |

Table 5.12 shows the weights $W$ and biases $B$ for the sixth layer in Fig. 5.12. The output of each neuron is given by (5.15), and Fig. 5.34 shows its hardware realization using multipliers implemented by SCMs.

$$L_{6,n} = f(In_1 W_{n,1} + In_2 W_{n,2} + In_3 W_{n,3} + In_4 W_{n,4} + B_n) \tag{5.15}$$

The whole ANN of six layers is shown in Fig. 5.35, at the output of each neuron, a register is located to avoid the delay through the logic path. The ANN is communicated to a PC using serial communication and the data listed in Table 5.13.

Figure 5.36 shows the hardware for the reception of data, where the enabled registers by *EOR* are activated each time a byte is received. When 4 bytes are received, *LHA* is activated and in the next clock cycle *LPA* is activated. This process is repeated until three packages are received, afterwards the registers enabled through *LGA* are activated, thus guaranteeing receiving data of three inputs at the same time. The output of the block *Blq: ANN* is obtained after 10 clock cycles.

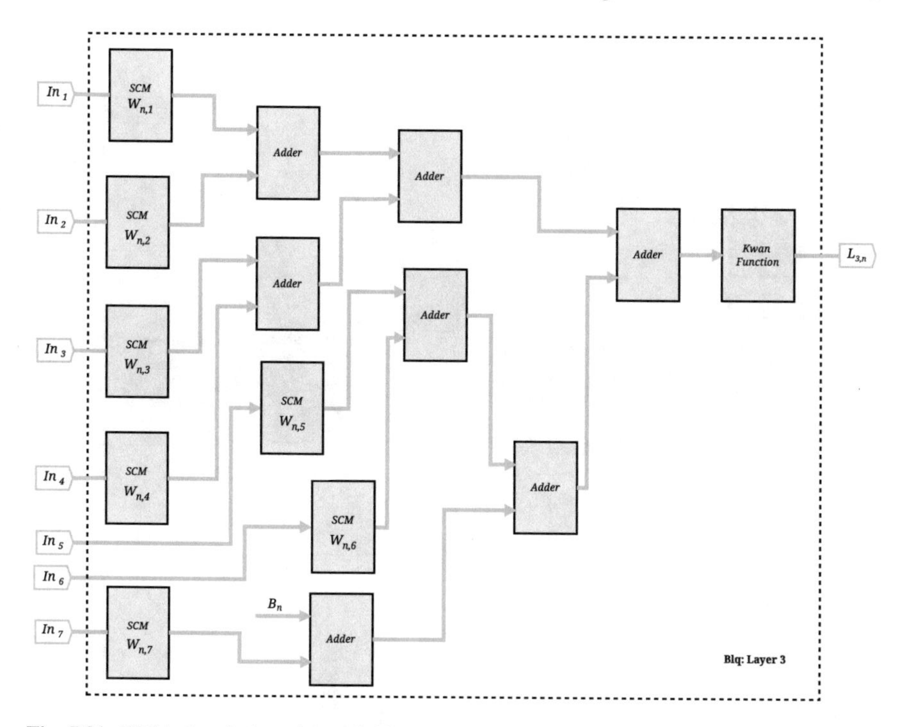


**Fig. 5.31** FPGA description of the third layer

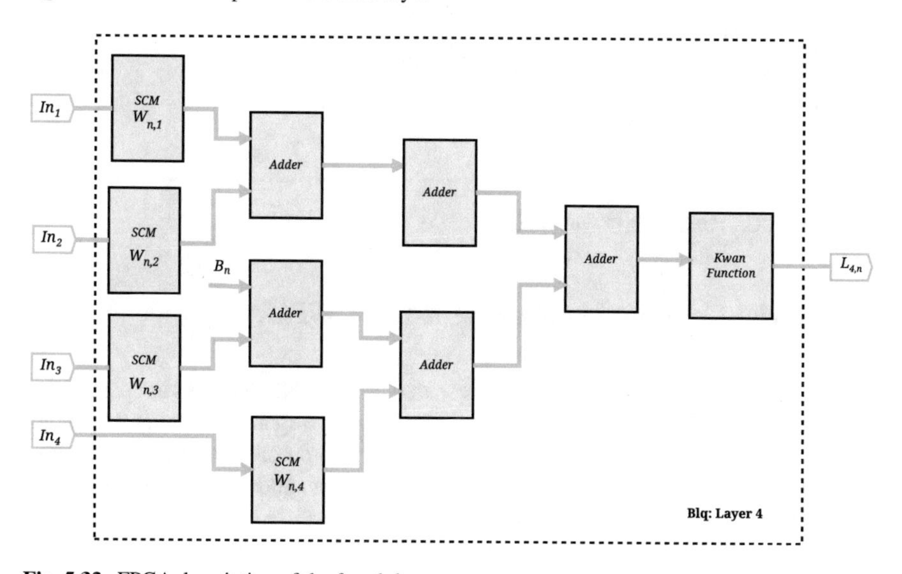


**Fig. 5.32** FPGA description of the fourth layer

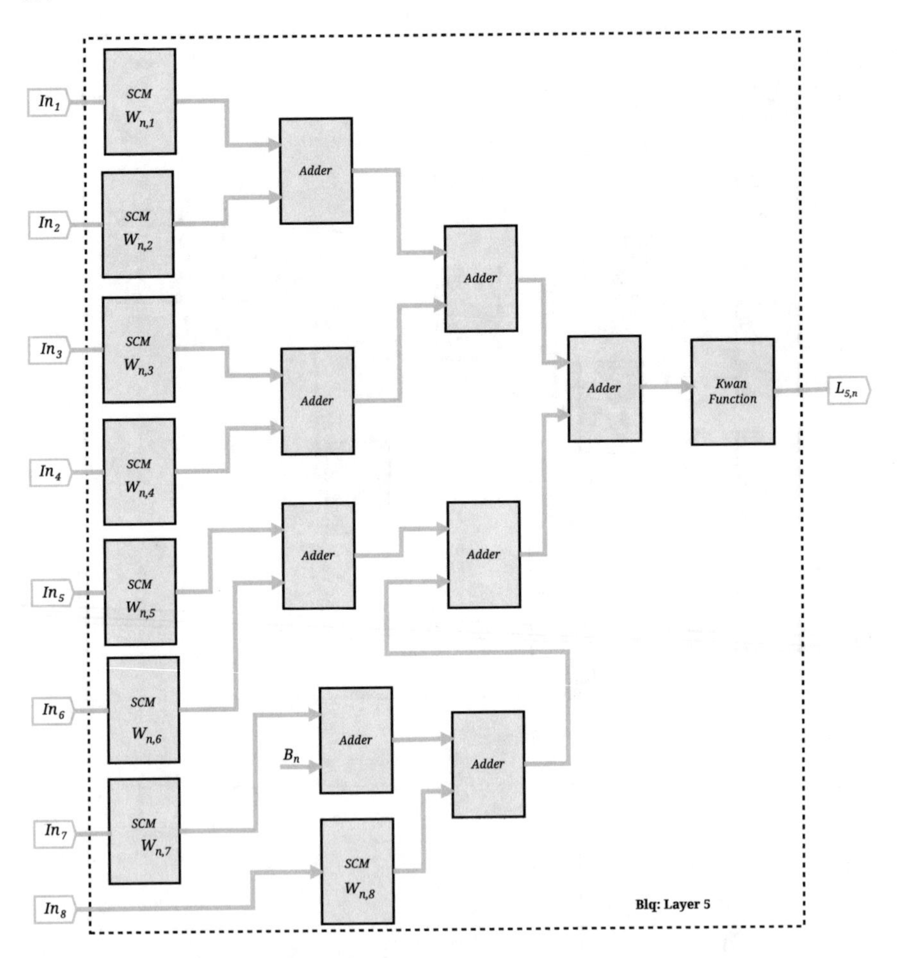


**Fig. 5.33** FPGA description of the fifth layer

## 5.5 Serial Communication Protocol: PC-FPGA

Soft computing represents an emergent research area, in which ANNs are part of this new computing field. ANNs process information and are organized based on the brain behavior [60]. They can operate in parallel and their interconnection pattern defines the architecture, training, learning, activation functions, and so on [61]. Extremely important is the selection of the number of neurons because using few neurons may lead to an under-fitting, so that the ANN may lack of resources to solve a problem. On the other hand, using many neurons may cause over-fitting [58]. That way, this is a challenge that is different for every problem at hand. Some authors recommend using one hidden layer and few neurons to solve practical problems, but still such a number depends on the complexity of the problem.

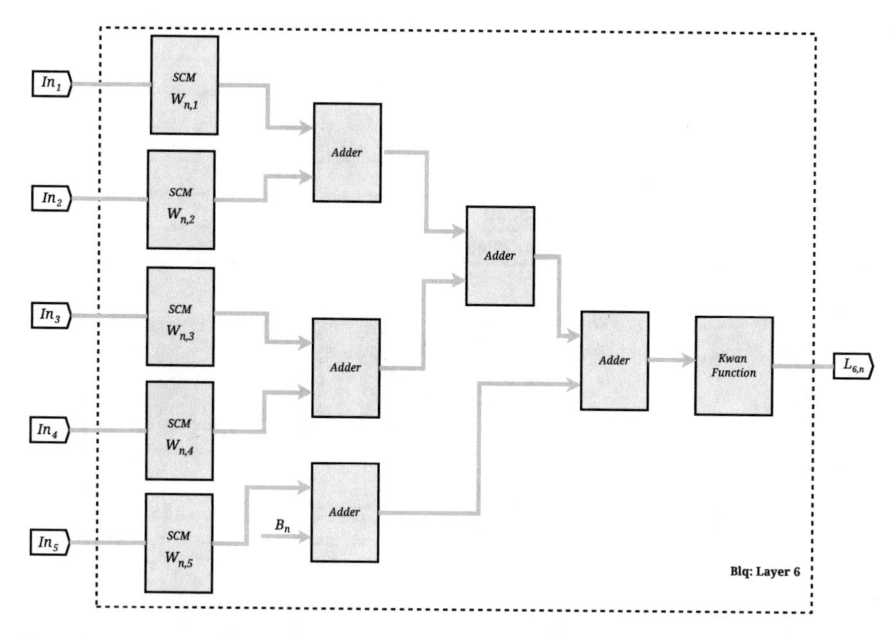


**Fig. 5.34** Hardware of the sixth layer in Fig. 5.12

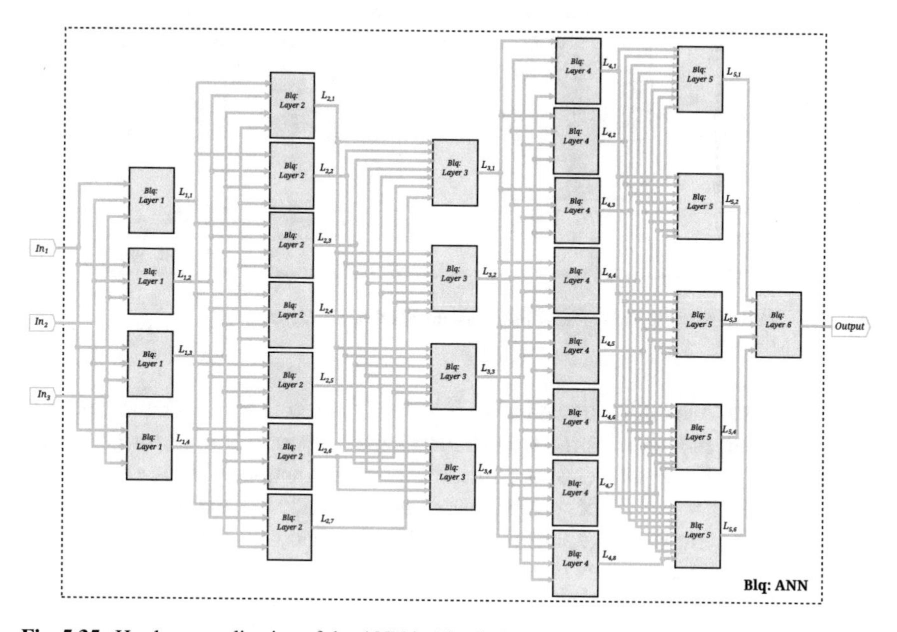


**Fig. 5.35** Hardware realization of the ANN in Fig. 5.12

**Table 5.13** Data: Altera Cyclone IV GX FPGA DE2i-150

| Component | Data |
|---|---|
| Oscillator clock input | 50 MHz |
| Baude rate | 115200 |
| Parity check bit | None |
| Data bits | 8 |
| Stop bits | 1 |
| Flow control (CTS/RTS) | ON |

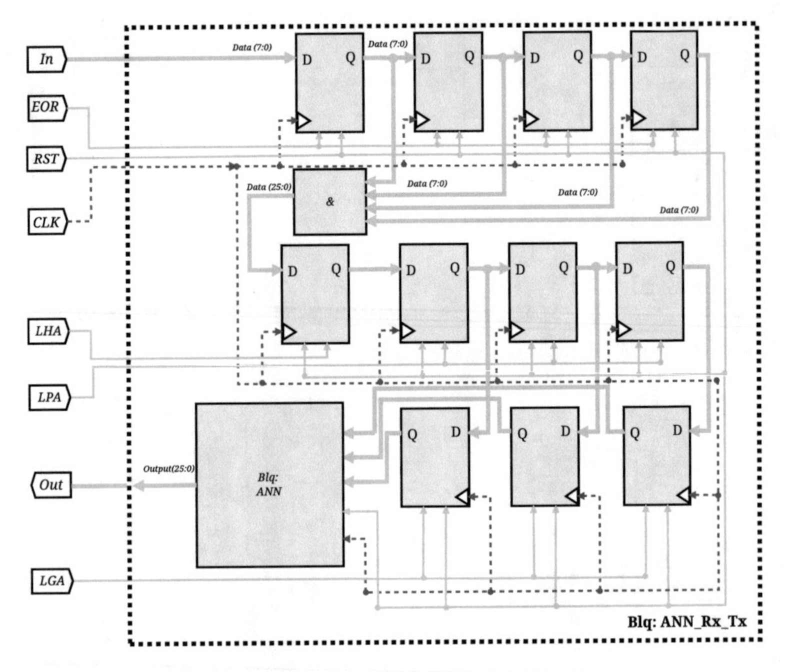


**Fig. 5.36** Hardware for the reception of data to the ANN of six layers

For instance, the ANN shown in Fig. 5.35 has $7 \times 4 \times 8 \times 5$ hidden layers. The input layer and hidden ones the hyperbolic tangent function is used for activation of the neurons. The last layer uses a linear activation function to reduce computing time. In the first layer, a TDL (tapped delay line) is placed so that the input signal (state variable $x$ from the time series generated by the chaotic oscillator) pass through $k-1$ delays (with $d=3$), as shown in Fig. 5.37.

As shown before, the hyperbolic tangent function is better if it is implemented by PWL approach [62]. With $\theta = 0.25$, $L = 2$ and $\beta = 1$, its block diagram is shown in Fig. 5.38.

Once the ANN is implemented as shown in Fig. 5.35, it can be communicated to a personal computer (PC) using the serial communication protocol [63]. This kind of communication involves the transmission of one bit at a time, where the total number

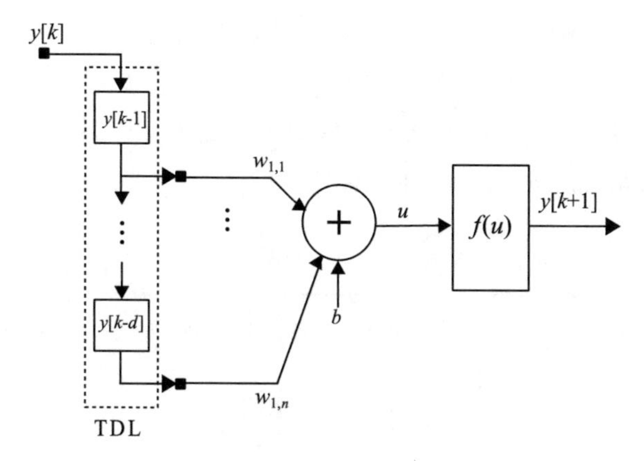


**Fig. 5.37** TDL associated to one neuron

of bits transmitted by second is named baud rate. From the data, each byte is divided into 8 bits and one by one is transmitted, which control is performed with other

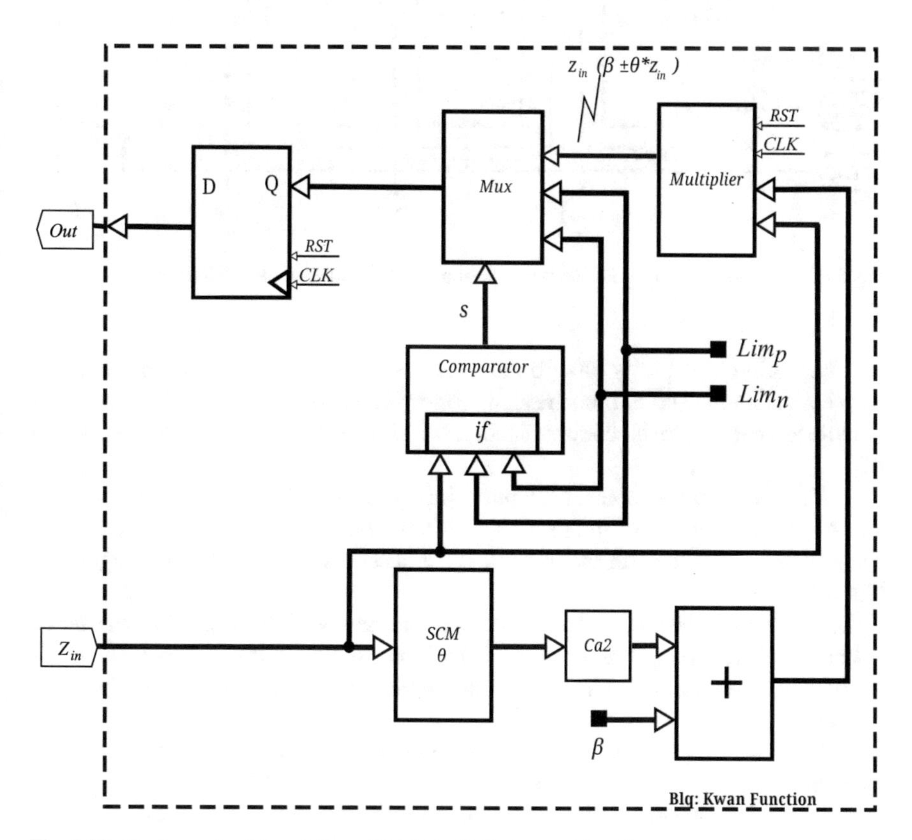


**Fig. 5.38** Hardware implementation of the hyperbolic tangent function

| | | | | | | | | | | | | | | | | | | | | | | | | | | | | | | | | |
|---|---|---|---|---|---|---|---|---|---|---|---|---|---|---|---|---|---|---|---|---|---|---|---|---|---|---|---|---|---|---|---|---|
| *Input₁* | 0 | 0 | 0 | 0 | 0 | 0 | 0 | 0 | 0 | 0 | 0 | 0 | 1 | 1 | 1 | 1 | 0 | 1 | 0 | 1 | 0 | 1 | 0 | 0 | 1 | 1 | 0 | 1 | 0 | 1 | 1 | 0 |
| *Input₂* | 0 | 0 | 0 | 0 | 0 | 0 | 0 | 0 | 0 | 0 | 0 | 0 | 0 | 1 | 1 | 1 | 0 | 1 | 0 | 1 | 1 | 1 | 0 | 0 | 1 | 1 | 0 | 1 | 1 | 1 | 1 | 1 |
| *Input₃* | 0 | 0 | 0 | 0 | 0 | 0 | 0 | 0 | 0 | 0 | 0 | 0 | 1 | 1 | 1 | 1 | 0 | 1 | 0 | 0 | 0 | 1 | 0 | 0 | 1 | 1 | 0 | 0 | 0 | 1 | 1 | 1 |

**Fig. 5.39** Distribution of the inputs of the ANN

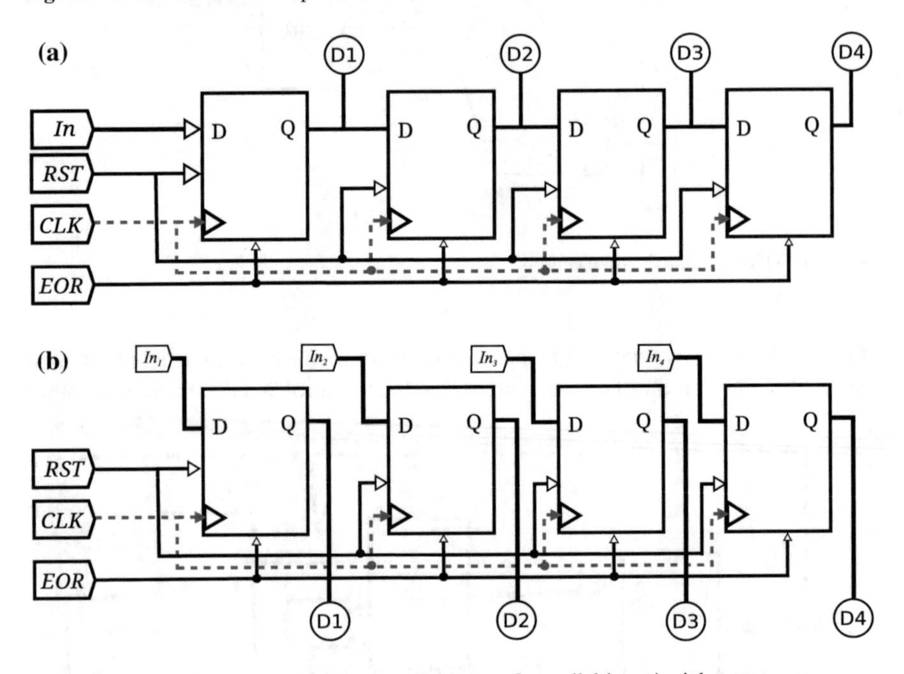


**Fig. 5.40** Shift registers: **a** serial-input/parallel-output, **b** parallel-input/serial-output

bits: beginning bits, parity bits (to detect errors during the transmission), and stop bits (indicating the end of transmission). Since just one transmitter and one receptor module are needed, serial communication is a simple way to connect an FPGA to a PC.

The ANN has three inputs, each one with a word length of 26 bits, so that 6 bits are concatenated as shown in Fig. 5.39. That way, each input will be updated to have a length of 32 bits; thus for the three inputs it is necessary to send 12 packages of 8 bits each one.

Due to its capacity of temporal storage and data shift by flip-flops, the block in charge of processing the received data will consist of shift registers with serial-input/parallel-output and other ones with parallel-input/serial-output, as shown in Fig. 5.40.

Figure 5.36 shows the block used to data reception, where the registers enabled by *EOR* will be activated each time a package is received. When four packages are received, the register enabled by *LHA* will be activated, in the next clock cycle *LPA* will be activated, and this process continues until 12 packages (the three inputs)

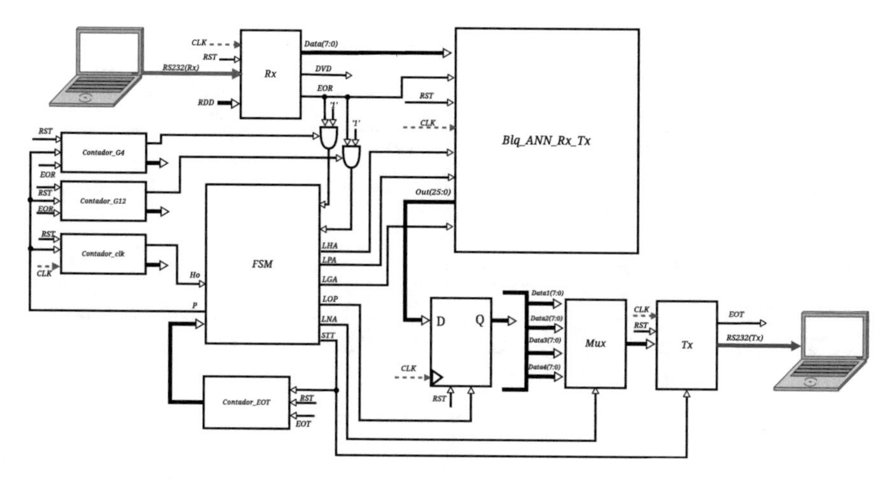


**Fig. 5.41** Serial communication between the FPGA and PC

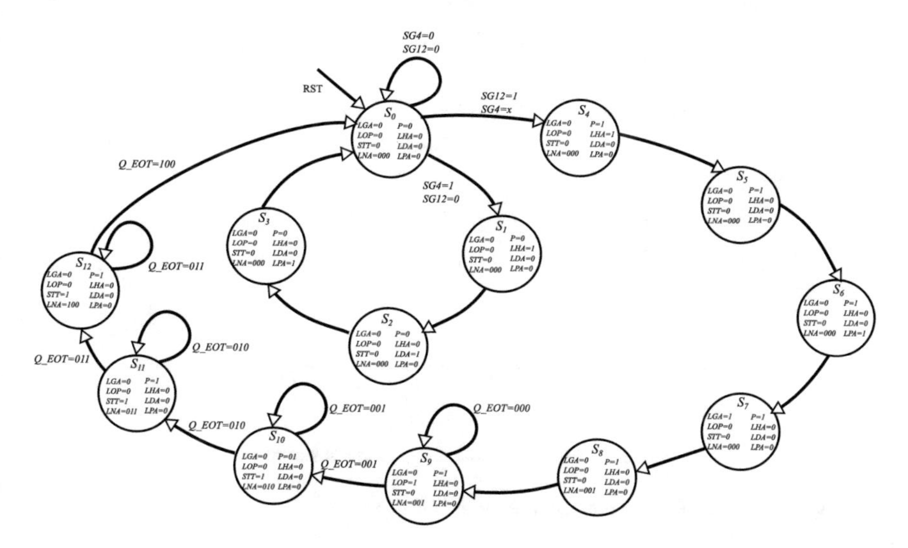


**Fig. 5.42** States machine for controlling the serial communication protocol

are received. Afterwards, the registers enabled by *LGA* will be activated, with this procedure one ensures receiving the data for the three inputs in the ANN at the same time. The output of the block Blq:ANN will be obtained after 10 clock cycles.

The block diagram sketching the transmission–reception for the serial communication between the FPGA and PC is shown in Fig. 5.41. As one sees, at the output of the block ANN-Rx-Tx a register is located that will be enabled after 10 clock cycles, which are required by the ANN. This data will be concatenated with 6 bits (to have a word of 32 bits, as shown in Fig. 5.39), so that four packages of 8 bits are created,

**Table 5.14** FPGA used resources for the ANN of six layers

| Resources | Available | Used | Percentage (%) |
|---|---|---|---|
| Logic elements | 146760 | 38428 | 26 |
| Combinatorial functions | 146760 | 38191 | 26 |
| Dedicated logic registers | 149760 | 1127 | <1 |
| Pins | 508 | 34 | 7 |
| 9 bits mutipliers | 720 | 196 | 27 |
| Fmax | | 26.32 MHz | |

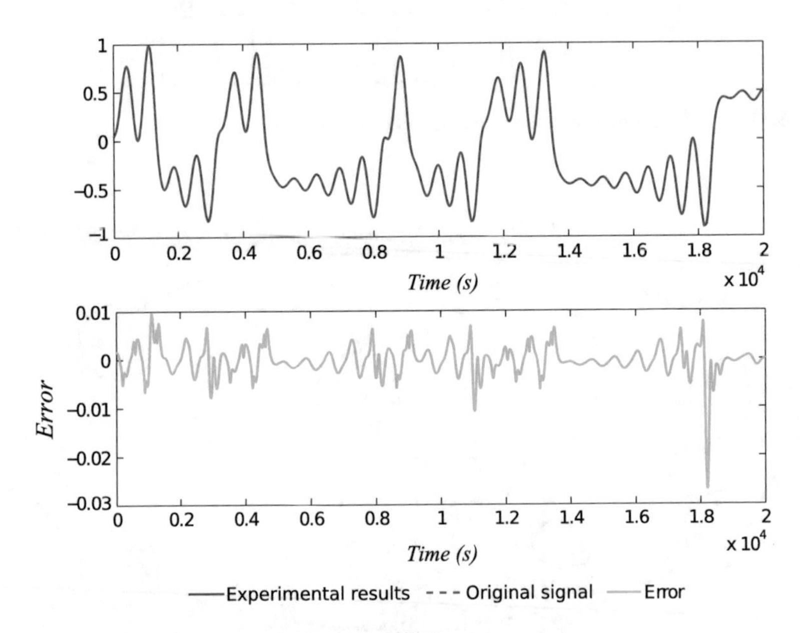


**Fig. 5.43** Chaotic time series prediction using ANNs

and which will be selected by a multiplexer when they are transmitted. Figure 5.42 shows the states machine to control the whole system from Fig. 5.41.

The utilized resources for the FPGA realization of the ANN are listed in Table 5.14. As one can see in Fig. 5.38, one multiplier is required while another can be implemented by an SMC. This reduces the maximum operating frequency (*FMax*) of the ANN, and then, taking as reference the FPGA clock, one must implement a frequency divider to obtain the required clock frequency of the ANN blocks. In this experiment MATLAB$^{TM}$ is in charge of the transmission and reception of data. The results are shown in Fig. 5.43, where one can observe an error that is due to the length of the digital word, which was selected from computer arithmetic. The error in this case increases during the negative transitions.

# Chapter 6
# Random Number Generators

## 6.1 Generating Pseudorandom Binary Sequences

From the logistic map equation:

$$y_{n+1} = 1 - ay_n^2, \tag{6.1}$$

with $0 < a \leq 2$. A binary sequence can be generated taking the threshold at 0:

$$b_{n+1} = \begin{cases} 0 & \text{if } y_{n+1} \leq 0, \\ 1 & \text{if } y_{n+1} > 0, \end{cases} \tag{6.2}$$

where $y_{n+1}$ is calculated with (6.1). This is a discrete system. Thus at every new value of $y_{n+1}$ a new binary number $b_{n+1}$, with 0 or 1 values, is generated. All the other chaotic systems that will be shown in Sect. 6.6 are continuous systems, and the binary sequence is taken by sampling the continuous signal, then it is necessary to calculate the sampling frequency, and also it will be necessary to calculate the threshold value (it is equal to zero in (6.2)).

The produced sequence is said to be "pseudo-random" because exactly the same sequence can be reproduced using the same initial conditions. In (6.2), the initial condition is a value for $y_0$. Truly random sequences, where it is impossible to generate the same sequence, is available in processes as playing a roulette, or throwing many times a fair coin.

## 6.2 Numerical Method for Solving a Chaotic Dynamical System

The Lorenz dynamical system and the Rossler one are modeled by three first order and ordinary differential equations (ODE). Also Chua's chaotic oscillator and the


E. Tlelo-Cuautle et al., *Engineering Applications of FPGAs*,
DOI 10.1007/978-3-319-34115-6_6

one based on saturated function series are modeled by systems of three first order ODEs, but also the oscillators are controlled by a function. That function consists of negative slopes in the case of Chua's chaotic oscillator, and of series of saturated functions for the other example that will be given in Sect. 6.6.

The point here is that it is necessary to use a numerical method to solve those chaotic systems to obtain the signals associated to the state variables.

Let an ODE of a single variable be defined as

$$\frac{df}{dx} = \dot{x} = f(x, \mathbf{a}), \tag{6.3}$$

where $f$ function depends on the independent state variable $x$ and, possibly several constants represented by vector $\mathbf{a}$. The simplest method to obtain $x$ in (6.3) is the Euler method that is given by

$$x_{i+1} = x_i + \Delta t \ f(x, \mathbf{a}), \tag{6.4}$$

which can be seen as the finite difference approximation of the derivative

$$f(x, \mathbf{a}) = \frac{x_{i+1} - x_i}{\Delta t}.$$

Euler integration method has an error proportional to $(\Delta t)^2$ but it is the simplest method to implement in hardware or in software, with the consideration that needs a very small integration time. For all the simulations in this chapter, $\Delta t = 0.001$ s was used to guarantee the stability of the numerical method, as discussed in Chap. 4.

## 6.3 Double-Scroll and Multi-scroll Chaos Generators

Chua's chaotic oscillator and the one based on a series of saturated functions, will be defined in this section. It will be defined also how to generate several scrolls with these kinds of oscillators.

### *6.3.1 Chua's Chaotic Oscillator*

The ODEs that model this chaotic oscillator [32, 40, 64] are

$$\begin{aligned} \dot{x} &= \alpha[y - x - g(x)], \\ \dot{y} &= x - y + z, \\ \dot{z} &= \beta y, \end{aligned} \tag{6.5}$$

where $\alpha$ and $\beta$ are two different constants, and $g(x)$ is the PWL function defined as

$$g(x) = m_{2n-1}x + \frac{1}{2}\sum_{i=k}^{2n-1}(m_{i-1} - m_i)(|x - b_i| - |x - b_i|), \tag{6.6}$$

where $m_i$ are slopes, that must be with negative values, and $b_i$ break-point values. The values given to function $g$ control the number of generated scrolls In order to generate an even number ($2n$) of scrolls, $k$ takes values $k = 1, 2, 3, \ldots$; to generate an odd number ($2n + 1$) of scrolls, $k$ takes values $k = 2, 3, 4, \ldots$.

The minimum number of scrolls is two, then $2n = 2$, $n = 1$, and $k = 1$. In this manner, (6.6) becomes

$$g(x) = m_1 x + \frac{1}{2}(m_0 - m_1)(|x + b_1| - |x - b_1|).$$

The graph of this example is shown in Fig. 6.1a for values $m_0 = -3.036$, $m_1 = -0.276$, and $b_1 = 0.1$. For generating 3-scrolls, $2n + 1 = 3$ then $n = 1$, and $k$ will take the values 2, and 3, and function $g(x)$ will be

$$g(x) = m_1 x + \frac{1}{2}(m_1 - m_2)(|x + b_2| - |x - b_2|) + \frac{1}{2}(m_2 - m_3)(|x + b_3| - |x - b_3|).$$

The graph is shown in Fig. 6.1b for values $m_2 = -3.036$, $m_1 = m3 = -0.276$, $b_1 = 0.8$, and $b_1 = 1.37$.

### 6.3.2 *Saturated Function Series-Based Chaotic Oscillator*

This chaotic oscillator can generate, as well as the Chua's one, more than 2-scrolls. It is described by the system of differential equations [31, 32, 65]

$$\begin{aligned} \dot{x} &= y \\ \dot{y} &= z \\ \dot{z} &= -ax - by - cz + d_1 f(x; m) \end{aligned} \tag{6.7}$$

where $a$, $b$, $c$, and $d_1$ are positive constants that can get values in the interval [0, 1]. The dynamical system is controlled by the PWL approximation, e.g., series of a saturated function $f$.

In the following, we describe in detail how the saturated function $f$ in (6.7) is obtained. Let $f_0$ be the saturated function

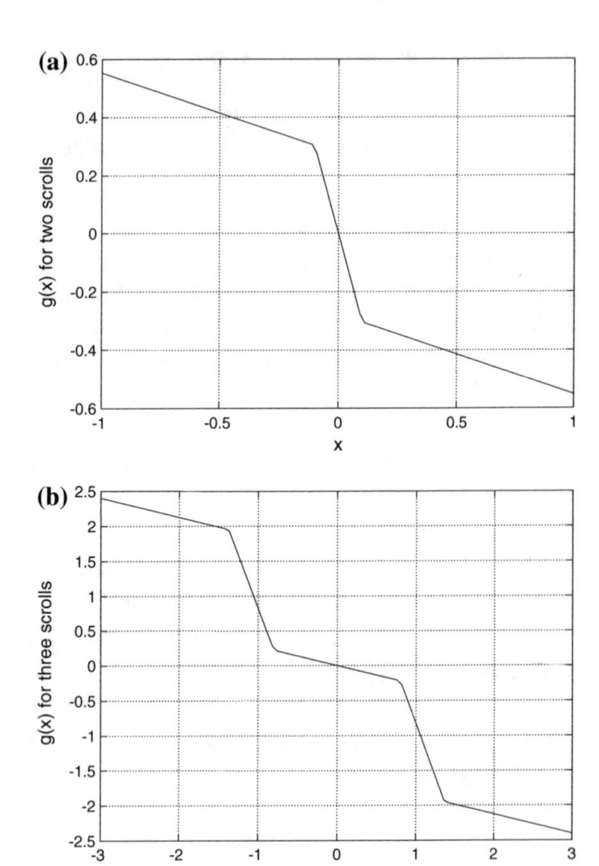


**Fig. 6.1** Graphs from (6.6) for generating 2-scrolls in **a**, and 3-scrolls in **b**. For both the graphs the following values were taken: $b_1 = 0.8$, $b_1 = 1.37$, $m_2$ of $m_3$ equal to $-3.036$, and $m_1 = -0.276$

$$f_0(x; m) = \begin{cases} 1, & \text{if } x > m \\ \frac{x}{m}, & \text{if } |x| \leq m \\ -1, & \text{if } x < -m, \end{cases} \tag{6.8}$$

where $1/m$ is the slope of the middle segment and $m > 0$, contrary to the negative slopes in Chua's oscillator; the upper radial $\{f_0(x; m) = 1 \; |x > m\}$, and the lower radial $\{f_0(x; m) = -1 \; |x < -m\}$ are called *saturated plateaus*, and the segment $\{f_0(x; m) = x/m \mid |x| \leq m\}$ between the two saturated plateaus is called *saturated slope*.

Let us now consider the saturated functions $f_h$ and $f_{-h}$ defined as:

$$f_h(x; m, h) = \begin{cases} 2, & \text{if } x > h + m \\ \frac{x-h}{m} + 1, & \text{if } |x - h| \leq m \\ 0, & \text{if } x < h - m, \end{cases} \tag{6.9}$$

and

$$f_{-h}(x;m,-h)=\begin{cases}0, & \text{if } x>h+m\\ \frac{x-h}{m}-1, & \text{if } |x-h|\le m\\ -2, & \text{if } x<h-m,\end{cases} \tag{6.10}$$

where $h$ is called the *saturated delay time* and $h > m$. Therefore, a saturated function series for a chaotic oscillator with $s$ scrolls is defined as the function

$$f(x;m)=\sum_{i=0}^{s-2} f_{2i-s+2}(x;m,2i-s+2) \tag{6.11}$$

where $s > 2$.

For example, using $f = f_0$ in (6.7), a 2-scrolls chaotic oscillator can be generated; the saturated function series for a 3-scrolls oscillator, $s = 3$, is generated from $i = \{0, 1\}$ in the sum in (6.11), then $f(x;m) = f_{-1}(x;m,-1)+f_1(x;m,1)$, and for a 4-scrolls attractor: $s = 4, i = \{0, 1, 2\}$ in (6.11), thus: $f(x;m) = f_{-2}(x;m,-2)+f_0(x;m)+f_2(x;m,2)$. Examples of $f$ function to generate 2- and 3-scrolls are shown in Chaps. 4 and 5.

## 6.4 Measuring the Entropy of a RNG

Entropy of a source $\mathscr{S} = \{S, P\}$, with a set of symbols $S = \{s_1, s_2, \ldots, s_n\}$ and probability distributions $P = \{p_1, p_2, \ldots, p_n\}$ is defined as [66, Chap. 15]

$$H(\mathscr{S})=-\sum_{i=1}^{n} p_i \log_2 p_i, \tag{6.12}$$

where $p_i = p(s_i)$, $p_i$ is the probability of success of symbol $s_i$.

If all the probabilities in (6.12) are equally probable, or the probability distribution is uniform, then (6.12) is reduced to

$$H(\mathscr{S})=-\log_2 p \tag{6.13}$$

where $p$ could be equal to any $p_i$ (all are equal).

We are going to visualize the meaning of (6.12) and (6.13) with an example. Suppose we have a fair coin with the option "head" coded as a 0, and the option "tail" coded as 1. As the coin is fair, both options have the probability $1/2$ of success, then its entropy will be

$$
\begin{aligned}
H &= -\frac{1}{2}\log_2\frac{1}{2} - \frac{1}{2}\log_2\frac{1}{2} \\
&= -\frac{1}{2}\left[2\log_2(2^{-1})\right] \\
&= \log_2(2) = 1
\end{aligned}
$$

The meaning of $H = 1$ is that we have a probability of $1/2$ to predict the next symbol given the predecessors. In fact, this is the maximum probability for the two symbols.

If the coin has a trick, i.e., it has two heads, the probability of this success is 1, then the entropy will be

$$H = -1\log_2(1) = -1(0) = 0.$$

Meaning in this case we can always predict a symbol (it will be always a head).

Linking now this measurement of entropy with a random number generator (RNG), we should expect that a good RNG produces sequences of 0/1 with an entropy equal to 1, meaning that every 0 or 1 has a probability of 1/2 of success; or in other words, in a good RNG should be impossible to predict the output of the next 0 or 1 given the previous ones. For a general event, a probability of 0.99 means that we are almost certain that such event will occur. If the probability is 0.001, we could think that the event will not occur. If the probability is 0.5 the uncertainty will be maximum [66, Chap. 15]. Then we expect an entropy of 1 for an unpredictable binary sequence. If the entropy of a RNG is less than 1, it means that the output could be predicted, with some method that is beyond the discussion of this chapter.

How the entropy can be measured in practice? A method is based on histograms [67, 68]: the occurrence of small sequences of two, three, or more binary symbols is computed. In [68] sequences of length 1 to 16 of a total of 1 million bits were used to calculate those histograms. The histograms form the estimated probabilities to calculate (6.12). The taken estimated entropy will converge to the true entropy as the number of taken sequences trend to infinity.

For all the entropy measurements taken in this chapter, the *context tree weighting* (CTW) algorithm was used [69]. The taken measurement by this algorithm is related with lossless compression schemes. The entropy bounds the performance of the strongest lossless compression, which can be realized in theory by using the typical set or in practice using Huffman, Lempel–Ziv or arithmetic coding schemes. Then, the performance of existing data compression algorithms is often used as a rough estimate of the entropy of a block of data [67, 69]. The study in [70] concludes that the CTW method is the most effective to measure the entropy, with the most accurate and reliable results.

## 6.5 NIST Measurements

The National Institute of Standards and Technology (NIST) of the United States of America had created a free software for testing pseudo and random number generators [(P)RNGs]. Specifically, NIST said that the test is for (P)RNGs for cryptographic applications, where randomness is a crucial characteristic. The package is available at,[1] and according to NIST [71, Chap. 5] it will address the problem of evaluating (P)RNGs for randomness. It will be useful in

- Identifying (P)RNGs which produce weak (or patterned) binary sequences,
- designing new (P)RNGs,
- verifying that the implementations of (P)RNGs are correct,
- studying (P)RNGs described in standards, and
- investigating the degree of randomness by currently used (P)RNGs.

All the instructions about how to compile and use the NIST package are in the Chap. 5 of Ref. [71]; in all the rest of this reference document are the full description of each of the 16 NIST tests and how to interpret their results. For each statistical test, a set of $p$-values (corresponding to the set of sequences) is produced. For a fixed significance level, a certain percentage of $p$-values are expected to indicate failure. For example, if the significance level is chosen to be 0.01 (i.e., $\alpha = 0.01$), then about 1 % of the sequences are expected to fail. A sequence passes a statistical test whenever the $p$-value $\geq \alpha$ and fails otherwise.

## 6.6 Different RNGs

In general, for each chaotic system or oscillator described by three ODEs like Chua's and the one based on saturated function series, the following steps were applied in this chapter, in order to generate random sequences:

1. A initial vector $[x_0, 0.0, 0.0]^{\mathrm{T}}$, with the value for $x_0$ is randomly chosen inside a given range.
2. A first output signal is generated using the Euler method for integrating the signal, in a step of $\Delta t = 0.001$ s. Always the values of the state variable $y$ were taken as the output signal.
3. 200 samples are taken of this first signal and his autocorrelation function is calculated. The first zero of the autocorrelated signal is taken as the sampling period to generate the binary samples. This idea was taken from [68].
4. In some of the chaotic systems was taken only the positive signals, because of the symmetry of the sampled signal taken in the previous step.
5. A histogram of a second sampled signal (perhaps only with the positive values) is calculated. The threshold value for generating the binary sequences was taken at the corresponding 0.5 value of the normalized and accumulated histogram.

[1] http://csrc.nist.gov/groups/ST/toolkit/rng/documentation_software.html.

6. 30 sequences of 5000 bits are generated and the entropy for each sequence is calculated. From these 30 results, the mean and standard deviation are calculated.

Three programs are involved in this process

1. For the generation of the first signal in the step 2.
2. A second program to generate the sampled signal in step 5.
3. A third program that generates the binary sequences in the last step.

For the simulation results in this section, all programs were coded in Python language.

For the five realizations in this section, the logistic map was given in Sect. 6.1, Chua's chaotic oscillator and the one based on saturated function series were explained in Sect. 6.3.2 and in Chap. 4. Only it is necessary to know two more realizations: the Lorenz and the Rossler systems. These both are systems of three differential equations. Lorenz system is expressed by

$$\begin{aligned}\dot{x} &= \sigma(y - x),\\ \dot{y} &= x(\rho - z) - y,\\ \dot{z} &= xy - \beta z.\end{aligned} \tag{6.14}$$

The Rossler system is defined by the following equations:

$$\begin{aligned}\dot{x} &= -y - z,\\ \dot{y} &= x + ay,\\ \dot{z} &= b + z(x - c).\end{aligned} \tag{6.15}$$

The parameters for all the implementations, and in order to generate a chaotic behavior, are given in Table 6.1. Visualization of the process for each implementation can be seen in Figs. 6.3, 6.4, 6.5, 6.6, 6.7, 6.8, 6.9 and 6.10.

**Table 6.1** Constants values in order to generate chaos

| Realization | List of values |
|---|---|
| Logistic map, Eq. (6.1) | $a = 1.8$ |
| Chua oscillator, Eq. (6.5) | $\alpha = 10$, $\beta = 15$ for 2-, 3- and 4-scrolls |
| 2-scrolls, in Fig. 6.3(a) | $m_0 = -3.036$, $m_1 = -0.276$, $b_1 = 0.1$ |
| 3-scrolls, in Fig. 6.4(a) | $m_2 = -3.036$, $m_1 = m_3 = -0.276$, $b_1 = 0.8$, $b_2 = 1.37$ |
| 4-scrolls, in Fig. 6.5(a) | $m_0 = m_2 = -3.036$, $m_1 = m_3 = -0.276$, $b_1 = 0.1$, $b_2 = 0.66$, $b_3 = 0.86$ |
| Saturated function series, Eq. (6.7) | |
| For 2-, 3- and 4-scrolls (in Figs. 6.6(a), 6.7(a), and 6.8(a), respectively) | $a = b = c = d_1 = 0.7$, and $m = 0.1$ |
| Lorenz system, Eq. (6.14), in Fig. 6.9(a) | $\sigma = 10$, $\rho = 28$, $\beta = 8/3$ |
| Rossler system, Eq. (6.15), in Fig. 6.10(a) | $a = 0.2$, $b = 0.2$, $c = 5.7$ |

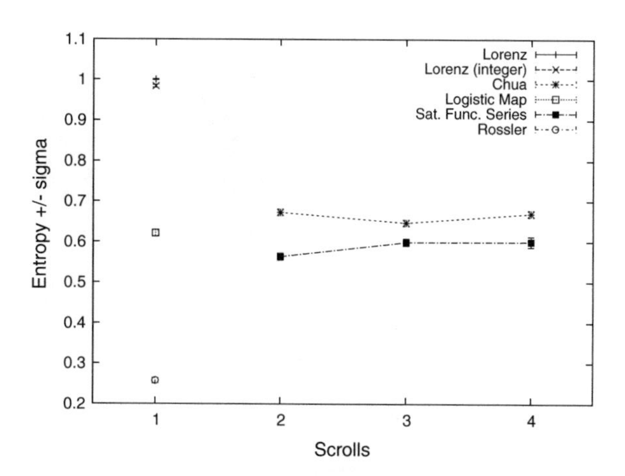


**Fig. 6.2** Mean ± standard deviation of 30 sequences of 5000 binary numbers of each realization

The final measurements of the entropy for each realization is shown in Fig. 6.2. The entropy for the logistic map is equal to the one obtained in [69], which confirms that our python implementation of the CTW algorithm works properly. The calculations for the Lorenz system were also performed with integer arithmetic with registers of 32 bits, showing a similar behavior that the implementation with real numbers (see Fig. 6.2).

The range of values for the initial $x_0$, the obtained sampling period, and the threshold value to generate the binary sequences are shown in Table 6.2. The value for $x_0$ is initialized with a random value within this given range in order to generate every sequence used to obtain the measurements in Fig. 6.2.

**Table 6.2** Values for the range of initial $x_0$, and the obtained sampling period and threshold to generate binary sequences for the different realizations

| Realization | $x_0$ range | Sampling period (s) | Threshold |
|---|---|---|---|
| Logistic map | [−1, 1] | 1 | 0 |
| Chua oscillator | | | |
| 2-scrolls (Fig. 6.3b–e) | [−0.2, 0.2] | 0.445 | 0.12 |
| 3-scrolls (Fig. 6.4b–e) | [−0.2, 0.2] | 0.445 | 0.288 |
| 4-scrolls (Fig. 6.5b–e) | [−0.2, 0.2] | 0.445 | 0.1177 |
| Sat. func. series | | | |
| 2-scrolls (Fig. 6.6b–e) | [−1.5, 1.5] | 1.95 | −0.025 |
| 3-scrolls (Fig. 6.7b–e) | [−1.5, 1.5] | 1.95 | −0.025 |
| 4-scrolls (Fig. 6.8b–e) | [−1.5, 1.5] | 1.95 | −0.025 |
| Lorenz system (Fig. 6.9b–e) | [−5, 5] | 0.726 | 0 |
| Rossler system (Fig. 6.10b–e) | [−5, 5] | 1.580 | 3.2 |

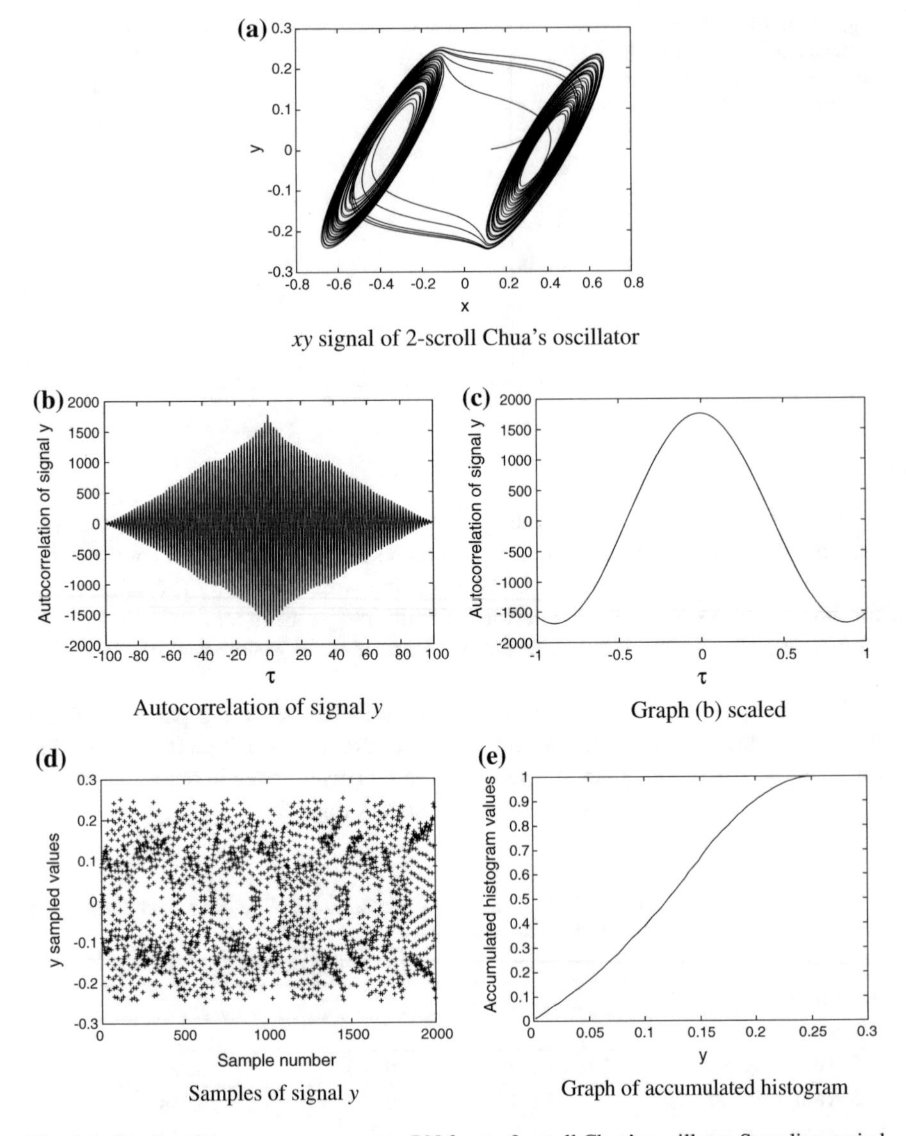


**Fig. 6.3** Graphs of the process to generate RN from a 2-scroll Chua's oscillator. Sampling period is taken from the signal crossing zero at graph **c**, and it is equal to 0.445 s. Because samples are symmetric as can be seen of graph **d**, only positive samples are taken. Threshold value is obtained from graph **e** at 0.5, then it is equal to 0.12

One million bits were generated with the Lorenz chaotic oscillator, Chua's 2-scroll oscillator, 3-scrolls saturated functions series-based oscillator, logistic map, and Rossler dynamical system. At least 100 sequences must be given to the NIST test suite, thus 10,000 (this is the $n$ value given to the test suite) is the length of each sequence. Five NIST tests were eliminated for testing the generated bits because

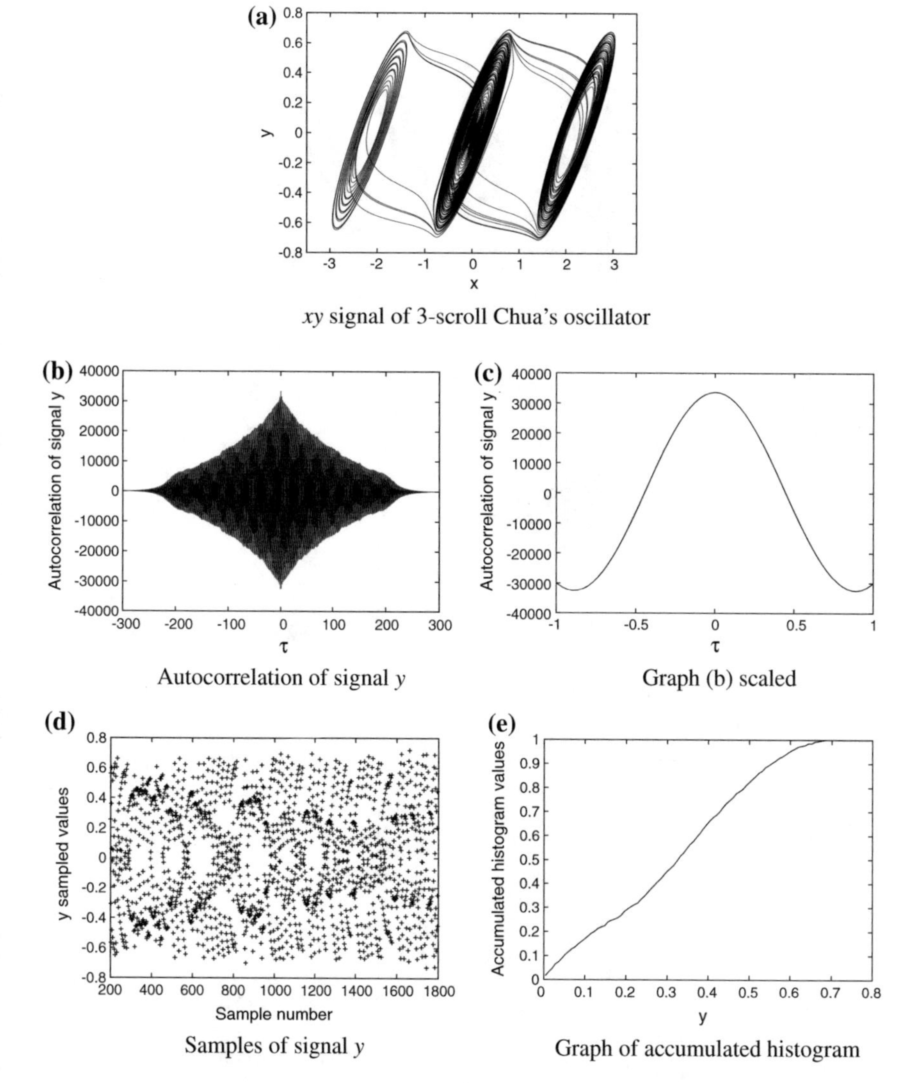


**Fig. 6.4** Graphs of the process to generate RN from a 3-scroll Chua's oscillator. Sampling period is taken from the signal crossing zero at graph **c**, and it is equal to 0.445 s. Because samples are symmetric as can be seen of graph **d**, only positive samples are taken. Threshold value is obtained from graph **e** at 0.5, then it is equal to 0.326

they need a bigger number of bits. The eliminated tests are *rank*, *random excursion*, *random excursion variant*, *universal*, *nonoverlapping templates*, and *overlapping templates*. In [72], the author used 160 sequences of 1 million bits each one in order to perform these five tests.

According to the documentation (pp. 2–30) [71], in order to run the NIST *approximate entropy* test, it is necessary to chose a value for $n$ and $m$, such that

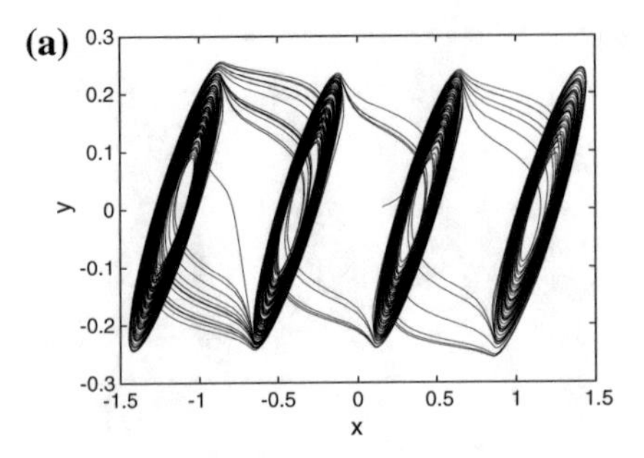


xy signal of 4-scroll Chua's oscillator

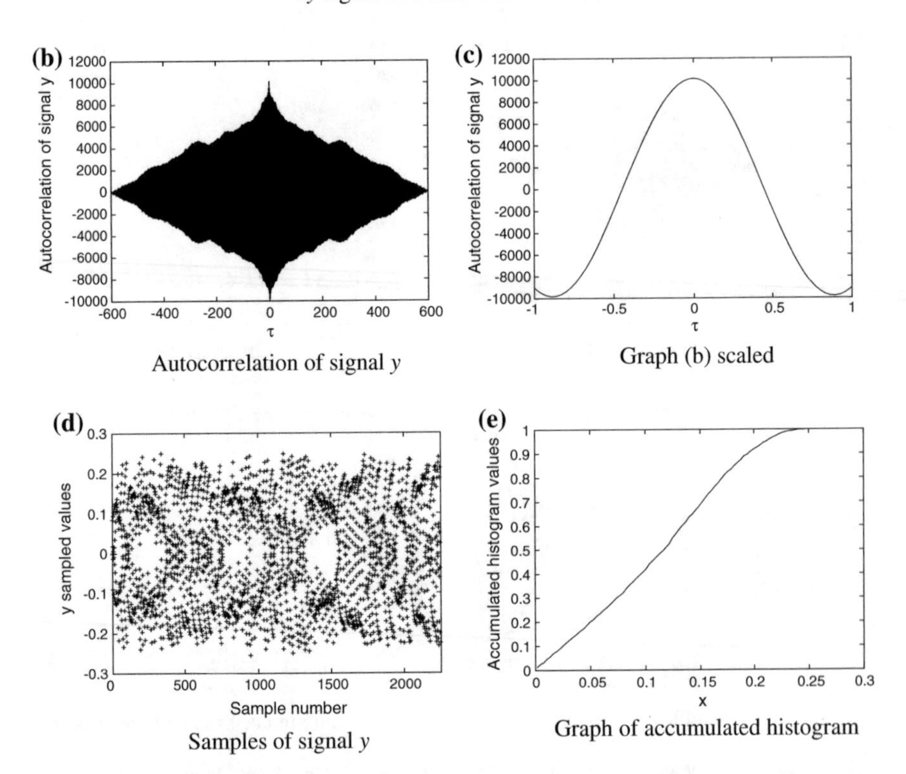


Autocorrelation of signal $y$

Graph (b) scaled

Samples of signal $y$

Graph of accumulated histogram

**Fig. 6.5** Graphs of the process to generate RN from a 4-scroll Chua's oscillator. Sampling period is taken from the signal crossing zero at graph **c**, and it is equal to 0.445 s. Because samples are symmetric as can be seen of graph **d**, only positive samples are taken. Threshold value is equal to 0.1177, this is the $x$ value obtained from graph **e** at 0.5 in the accumulated histogram

$m < \lfloor \log_2 n \rfloor - 5$, therefore, for the calculations performed herein: $n = 10{,}000$, and $m = 7$.

For the *serial* test (pp. 2–28) [71], values for $n$ and $m$ must satisfy $m < \lfloor \log_2 n \rfloor - 2$, thus, for $n = 10{,}000$, $m$ was selected equal to 10.

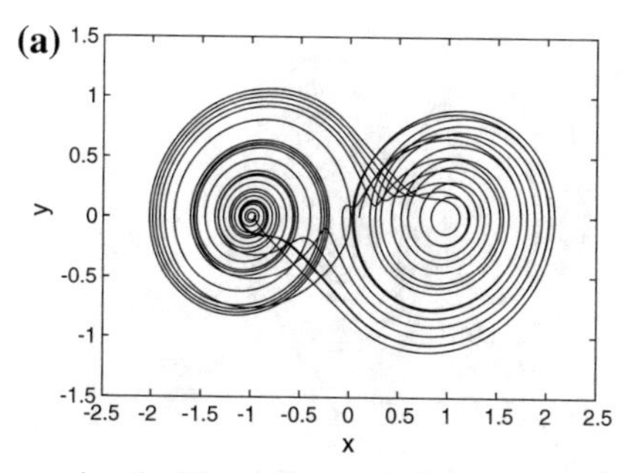


*xy* signal of 2-scroll saturated function series based oscillator

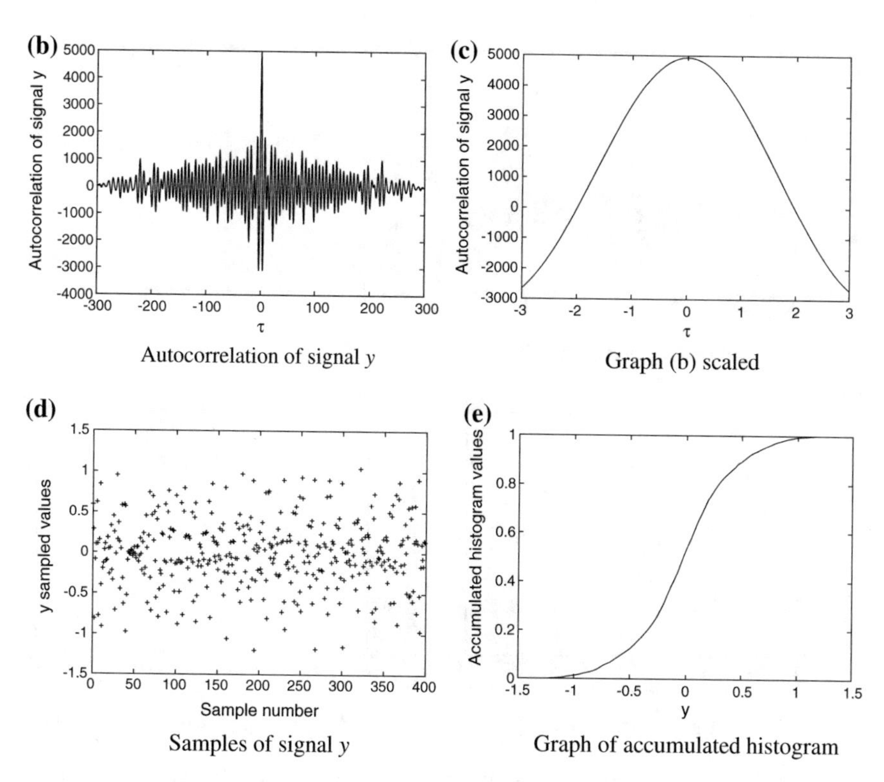


**Fig. 6.6** Graphs of the process to generate RN from a 2-scroll saturated function series-based oscillator. Sampling period is taken from the signal crossing zero at graph **c**, and it is equal to 1.95 s. Samples of signal $y$ in **d**. Threshold value is equal to $-0.025$ taken at 0.5 in the accumulated histogram in graph **e**

Results for the rest of nine tests are shown in Table 6.3. The *cumulative*, and the *serial* tests generate two values each one, therefore, there are 11 values per column in Table 6.3. To pass a test the $p$-value must be greater than 0.01, and the proportion

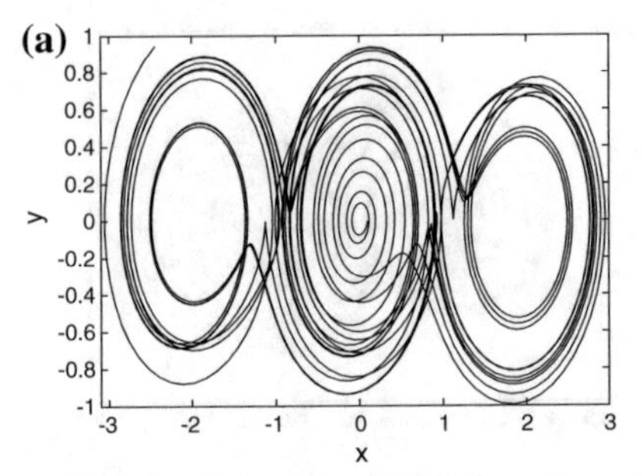


*xy* signal of 3-scroll saturated function series based oscillator

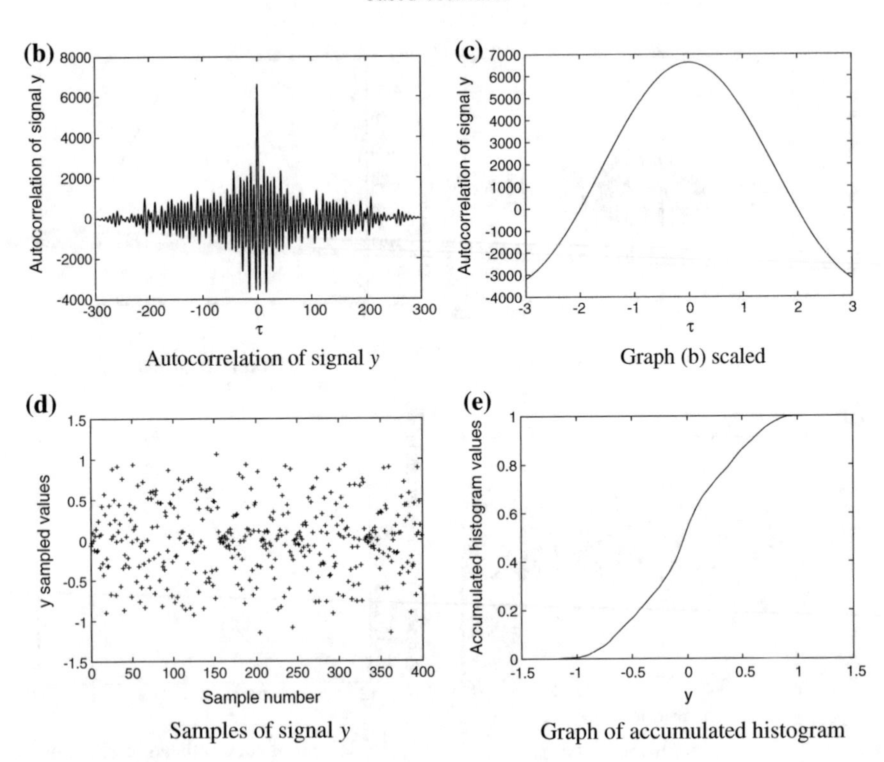


Autocorrelation of signal *y*

Graph (b) scaled

Samples of signal *y*

Graph of accumulated histogram

**Fig. 6.7** Graphs of the process to generate RN from a 3-scroll saturated function series-based oscillator. Sampling period is taken from the signal crossing zero at graph **c**, and it is equal to 1.95 s. Samples of signal *y* in **d**. Threshold value is equal to −0.025 taken at 0.5 in the accumulated histogram in graph **e**

value must be greater or equal than 0.96. This value of 0.01 is the default in the NIST test (must be between 0.01 and 0.001), and the 0.96 value is given in the output file. Serial test fail for the bits generated from the Lorenz system because the first value passes the test, but not the second.

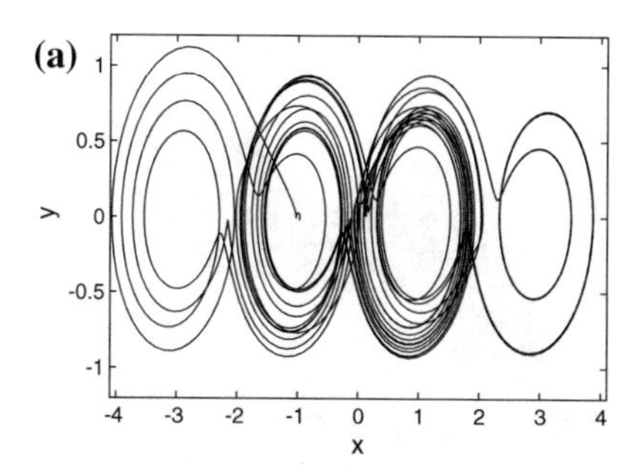


*xy* signal of 4-scroll saturated function series based oscillator

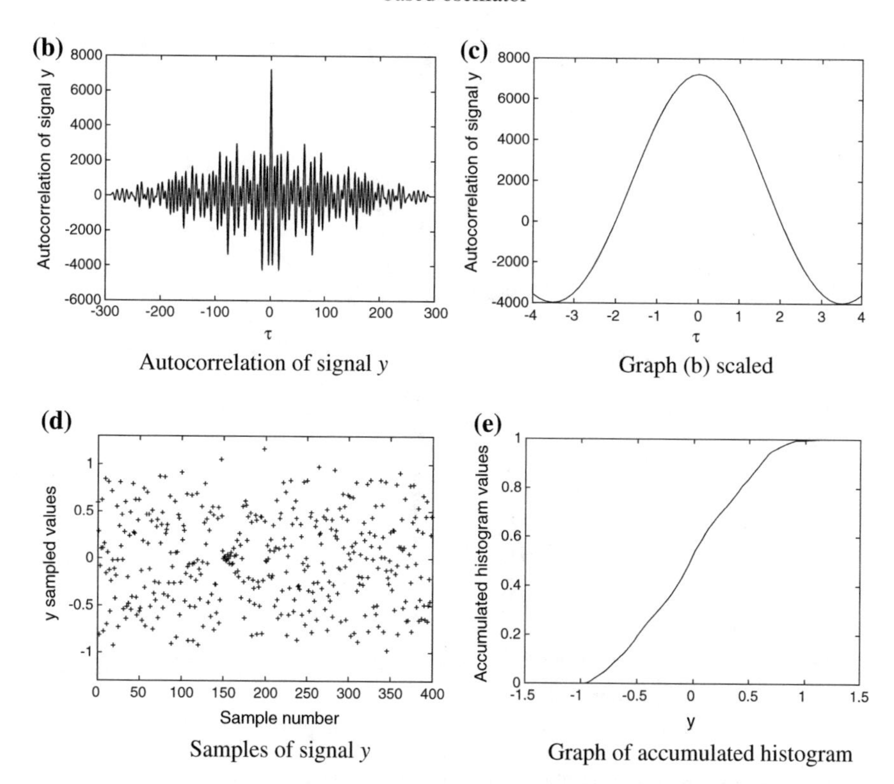


**Fig. 6.8** Graphs of the process to generate RN from a 4-scroll saturated function series-based oscillator. Sampling period is taken from the signal crossing zero at graph **c**, and it is equal to 1.95 s. Samples of signal $y$ in **d**. Threshold value is equal to $-0.025$ taken at 0.5 in the accumulated histogram in graph **e**

In Table 6.3, it is included for comparison the NIST test applied to one million bits generated with the Python module *random*: this sequence passes all the nine tests. The binary sequence generated with Lorenz system had an entropy almost of 1.0 (see

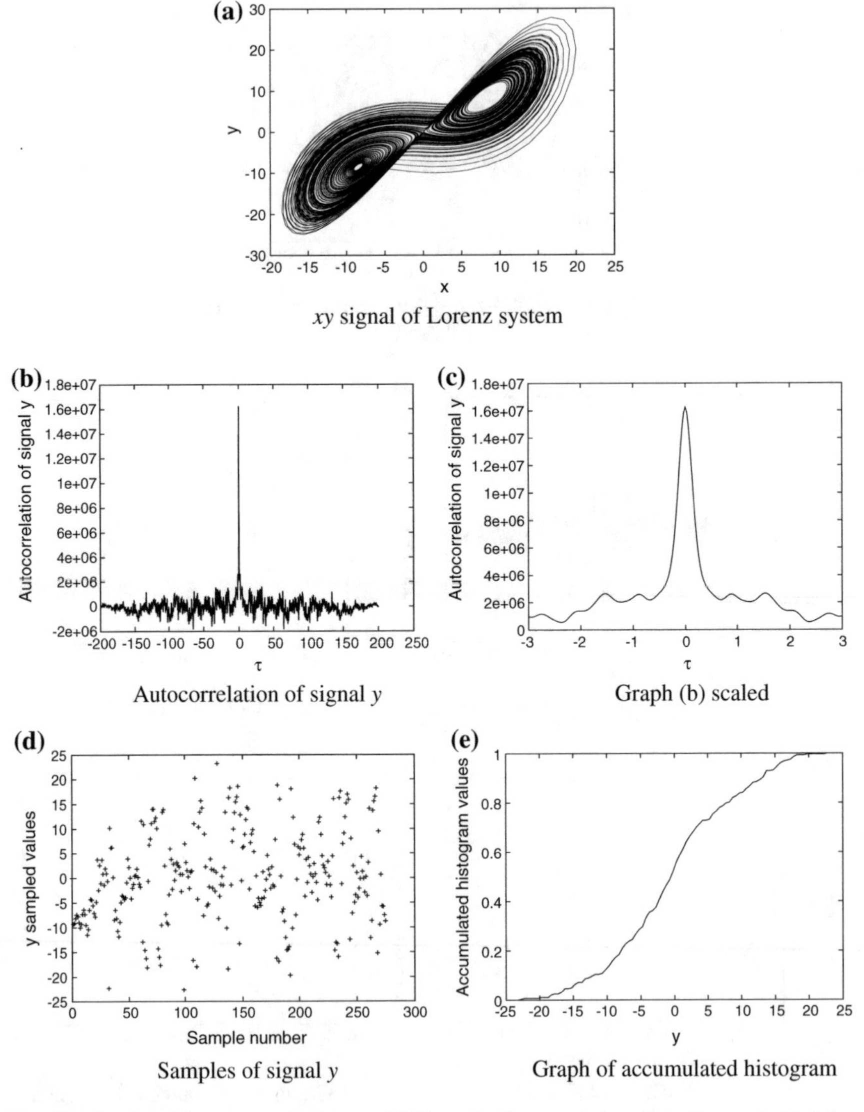


**Fig. 6.9** Graphs of the process to generate RN from the Lorenz system. Sampling period is taken at 0.726 in graph **c**. Samples of signal $y$ in **d**. Threshold value is equal to 0 taken at 0.5 in the accumulated histogram in graph **e**

Fig. 6.2) and passed four tests. The four rest of sequences in Table 6.3, generated with Chua's oscillator with 2-scrolls, saturated function series (SFS)-based oscillator with 3-scrolls, logistic map, and Rossler system, only passed one test: the *linear complexity* test, which could mean that all sequences are generated with chaotic realization.

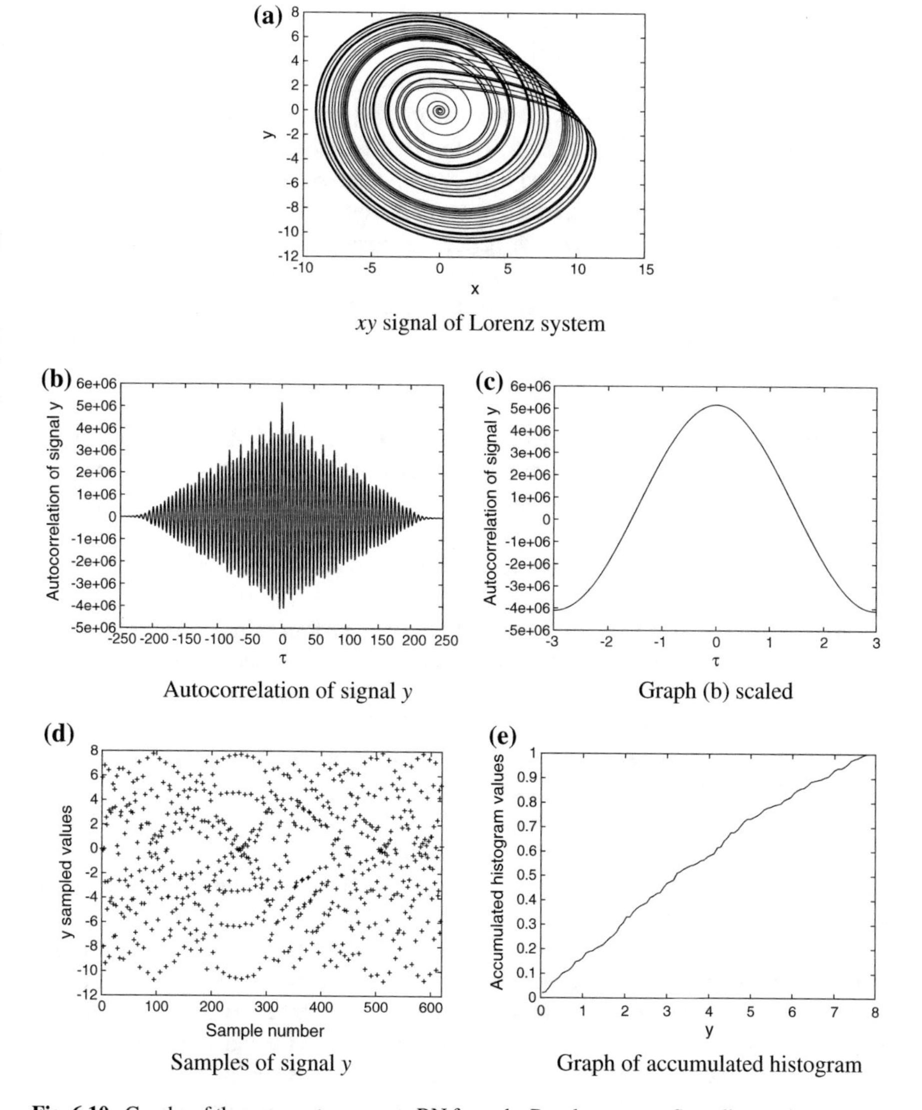


**Fig. 6.10** Graphs of the process to generate RN from the Rossler system. Sampling period is taken at 1.580 in graph **c**. Samples of signal $y$ in **d**. Threshold value is equal to 3.2 taken at 0.5 in the accumulated histogram in graph **e**

The interpretation of results in Table 6.3 could be that, with the methodology proposed in this chapter, it is not possible to generate a truly random binary sequence. In [73], the authors generated truly random binary sequences from a 2-scroll chaotic oscillator, the characteristic that authors used in order to determine a random bit is the transition between the scrolls. In [72], the authors examine the distribution of $x$

values in the stroboscopic Poincaré maps of their chaotic system, and in this it was the one based on the sampling period.

Although, according to data in Table 6.3, the generated sequences are not definitely random, this kind of sequences were used in [74] into the genetic operators for Evolutionary Algorithms. According to the results of authors in [74], results are better by using sequences generated with chaos, in the same way that was explained in this chapter. This result could lead us that randomness could not be very important, and, this is only a hypothesis, could be better sequences that cover all the search space, such as Halton [75] sequences do.

NIST test source code is in C language. Once it is compiled, using gcc, its execution can be performed as

```
./assess 10000 < input.in
```

where the file `input.in` has content shown in Fig. 6.11; `10000` is the number of bits per sequence, and the file with the generated zeros and ones is in the location `data/ros.rnd`. In this way it is possible to use `assess` program on the command line, noninteractively.

As one can see in Table 6.3 none of the generated sequences passed all the NIST test.

The sequence generated with Lorenz system passes four of the nine tests. A new sequence was generated by applying the *bit counting redundancy reduction technique* used in [76]: The original Lorenz sequence is divided in blocks of 5 bits, a new bit for a new sequence is generated applying the XOR operation on the 5 bits of each block. This new sequence passed all nine NIST test (see Table 6.4). Then it could be considered as a pseudorandom sequence.

Applying the bit counting redundancy reduction technique with 6 bits, to the original sequence produced by the logistic map, increases the NIST passed tests from 1 to 2 (see Tables 6.3 and 6.4). But this new sequence cannot be considered a pseudorandom sequence.

The von Neumann de-skewing technique [73] was also applied to the original logistic map sequence. This de-skewing technique eliminates the correlation in the output of the natural sources of random bits: pair of independent 01 bits are converted to 0, and in the same way 10 bits are converted to 1, pairs 00 and 11 are discharged. This new sequence passes one more NIST test, as can be seen in Table 6.4, and this new sequence also cannot be considered as a pseudorandom sequence.

In [73], a true random sequence was generated by detecting the jumps of the $x$ signal between 2-scrolls in a double-scroll chaotic oscillator. A similar technique is used here using the Chua's 2-scroll oscillator: A binary sequence is generated using (6.2), but now using the signal $x$. Then this binary sequence is processed as

$$c_i = \begin{cases} 0 & \text{if } b_{2i-1} = 0 \text{ and } b_{2i} = 1, \\ 1 & \text{if } b_{2i-1} = 1 \text{ and } b_{2i} = 0, \end{cases} \tag{6.16}$$

**Table 6.3** Results of apply NIST tests to some generated sequences

| | Python random function | | Lorenz | | Chua 2s | |
|---|---|---|---|---|---|---|
| | *p*-value | Proportion | *p*-value | Proportion | *p*-value | Proportion |
| Frequency | 0.319084 | 0.98 | 0.514124 | 0.98 | 0.000000 | 0.11 |
| Block frequency | 0.911413 | 1.00 | 0.000000 | 0.80 | 0.000000 | 0.09 |
| Cumulative sum forward | 0.162606 | 0.98 | 0.122325 | 0.99 | 0.000000 | 0.12 |
| Cumulative sum reverse | 0.897763 | 0.98 | 0.108791 | 0.99 | 0.000000 | 0.07 |
| Runs | 0.102526 | 1.00 | 0.000000 | 0.00 | 0.000000 | 0.00 |
| Longest run of ones | 0.699313 | 0.98 | 0.000000 | 0.49 | 0.000000 | 0.00 |
| Spectral DFT | 0.759756 | 0.99 | 0.275709 | 1.00 | 0.000000 | 0.00 |
| Approximate entropy | 0.437274 | 0.98 | 0.000000 | 0.01 | 0.000000 | 0.00 |
| Linear complexity | 0.779188 | 0.97 | 0.171867 | 0.96 | 0.202268 | 0.94 |
| Serial | 0.678686 | 0.99 | 0.000000 | 0.19 | 0.000000 | 0.00 |
| | 0.334538 | 0.99 | 0.075719 | 1.00 | 0.000000 | 0.00 |
| Total tests passed | 9 | | 4 | | 1 | |

| | SFS 3-scrolls | | Log. map | | Rossler | |
|---|---|---|---|---|---|---|
| | *p*-value | Proportion | *p*-value | Proportion | *p*-value | Proportion |
| Frequency | 0.000000 | 0.00 | 0.000000 | 0.00 | 0.000000 | 0.00 |
| Block frequency | 0.000000 | 1.00 | 0.000000 | 0.00 | 0.000000 | 0.00 |
| Cumulative sum forward | 0.000000 | 0.00 | 0.000000 | 0.00 | 0.000000 | 0.00 |
| Cumulative sum reverse | 0.000000 | 0.00 | 0.000000 | 0.00 | 0.000000 | 0.00 |
| Runs | 0.000000 | 0.00 | 0.000000 | 0.00 | 0.000000 | 0.00 |
| Longest run of ones | 0.000000 | 0.00 | 0.000000 | 0.87 | 0.000000 | 0.00 |
| Spectral DFT | 0.000000 | 0.00 | 0.000000 | 0.00 | 0.000000 | 0.00 |
| Approximate entropy | 0.000000 | 0.00 | 0.000000 | 0.00 | 0.000000 | 0.00 |
| Linear complexity | 0.834308 | 0.97 | 0.437274 | 1.00 | 0.419021 | 0.97 |
| Serial | 0.000000 | 0.00 | 0.000000 | 0.00 | 0.000000 | 0.00 |
| | 0.000000 | 0.00 | 0.000000 | 0.00 | 0.000000 | 0.00 |
| Total tests passed | 1 | | 1 | | 1 | |

**Table 6.4** Pseudorandom sequences, according to the shown NIST tests, are generated with Lorenz system plus the 5 bit counting technique, and with the Chua's 2-scrolls oscillator following (6.16), plus the same bit counting technique

| | Lorenz + 5 bit counting | | Log. map + 6 bit counting | | Log. map + von Neumann | |
|---|---|---|---|---|---|---|
| | $p$-value | Proportion | $p$-value | Proportion | $p$-value | Proportion |
| Frequency | 0.911413 | 1.00 | 0.000000 | 0.24 | 0.213309 | 1.00 |
| Block frequency | 0.514124 | 0.99 | 0.000000 | 0.91 | 0.000000 | 1.00 |
| Cumulative sum forward | 0.249284 | 1.00 | 0.000000 | 0.25 | 0.000000 | 1.00 |
| Cumulative sum reverse | 0.657933 | 1.00 | 0.000000 | 0.28 | 0.000000 | 1.00 |
| Runs | 0.304126 | 1.00 | 0.000000 | 0.67 | 0.000000 | 0.00 |
| Longest run of ones | 0.924076 | 1.00 | 0.000000 | 0.93 | 0.000000 | 0.00 |
| Spectral DFT | 0.181557 | 0.98 | 0.236810 | 1.00 | 0.000000 | 0.06 |
| Approximate entropy | 0.759756 | 0.98 | 0.000000 | 0.93 | 0.000000 | 0.00 |
| Linear complexity | 0.964295 | 0.99 | 0.616305 | 0.98 | 0.911413 | 0.96 |
| Serial | 0.289667 | 0.99 | 0.000022 | 0.98 | 0.000000 | 0.00 |
| | 0.058984 | 0.96 | 0.474986 | 0.99 | 0.000000 | 0.00 |
| Total tests passed | 9 | | 2 | | 2 | |

| | Chua 2-scrolls $x$ | |
|---|---|---|
| | $p$-value | Proportion |
| Frequency | 0.779188 | 1.00 |
| Block frequency | 0.739918 | 0.98 |
| Cumulative sum forward | 0.719747 | 1.00 |
| Cumulative sum reverse | 0.574903 | 0.99 |
| Runs | 0.383827 | 0.98 |
| Longest run of ones | 0.275709 | 0.97 |
| Spectral DFT | 0.016717 | 0.98 |
| Approximate entropy | 0.032923 | 0.98 |
| Linear complexity | 0.066882 | 0.96 |
| Serial | 0.867692 | 0.98 |
| | 0.924076 | 1.00 |
| Total tests passed | 9 | |

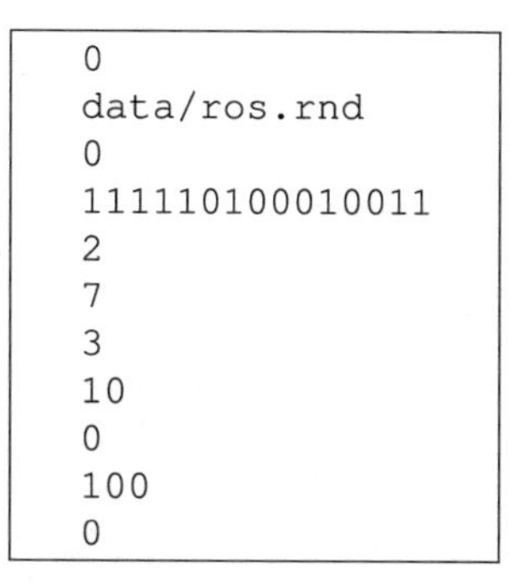

```
0
data/ros.rnd
0
111110100010011
2
7
3
10
0
100
0
```

**Fig. 6.11** Contents of `input.in` file to avoid to use interactively program `assess`

for $i = 1, 2, \ldots,$ and samples of sequence $b$ are taken every 12 integration steps. The sequence $c_i$ is generated when two independent samples cross the threshold of 0 in the $x$ variable. This new sequence passes all the nine NIST tests as can be seen in Table 6.4. The generation of this sequences takes a lot of time compared with all the other sequences. This is because in all the other sequences, a bit is generated each 1, 2, or 5 times the sampling period (2 samples are necessary to apply von Neumann de-skewing technique, or when only the positive samples are taken, and 5 samples are necessary if the 5 bit counting redundancy reduction technique is applied). Which is the average time that takes a bit to be generated with (6.16), of how to increase the throughput of this kind of pseudorandom sequences is a future work.

A final note for this chapter. The property of *noninvertibility* of the function that generated the random sequences is very important. This topic is also related to *one way functions*: It should be impossible to predict a new bit of the sequences knowing all the previous bits. Appears to be that this property is fulfilled with three degree dynamical systems if only the signal of a single variable is taken [72, 73].

# Chapter 7
# Secure Communication System

## 7.1 Chaotic Secure Communication Systems

Chaos and its applications in the field of secure communication was partially motivated by the fact that power spectrums of chaotic systems resemble white noise; thus making them an ideal choice for carrying and hiding signals over the communication channel [79]. The challenge in designing chaos-based secure communication systems can be stated as how to send a secret message from the transmitter (drive system) to the receiver (response system) over a public channel while achieving security, maintaining privacy, and providing good noise rejection.

In [78], one can find a general summary of major chaos-based modulation methods being investigated and developed internationally for communication applications. They are the following:

- Additive chaotic masking. This was the earliest form of modulation, wherein the information is added to the carrier as a small perturbation and usually demodulated using a cascaded form of master–slave synchronization.
- Chaotic switching. In this chaos-based version of traditional digital modulation, an analog signal of finite duration represents a digital symbol consisting of one or more bits. In this case, the digital symbol is uniquely mapped to an analog waveform segment coming from distinct strange attractors, or an analog waveform segment from a distinct region of a single strange attractor, thereby forming a chaotic signal constellation.
- Forcing function modulation. In this approach, a sinusoidal forcing function in a nonautonomous chaotic system is analog or digitally modulated with the information in a classical manner, with the transmitted signal being some other state variable. This modulation typically involves the nonautonomous or inverse synchronization methods and is the basis for the Aerospace development effort addressing high-data-rate, chaos-based communications.
- Multiplicative chaotic mixing. This can be considered as the chaos-based version of the traditional direct-sequence spread-spectrum approach, except in this case, the receiver actually divides by the chaotic carrier to extract the original information.


E. Tlelo-Cuautle et al., *Engineering Applications of FPGAs*,
DOI 10.1007/978-3-319-34115-6_7

- Parametric modulation. In this case, the information directly modulates a circuit parameter value (such as resistance, capacitance, or inductance), and some state variable from the chaotic system is sent that contains the information in a complex manner. As with forcing function modulation, this is an indirect modulation approach that typically offers higher levels of privacy and security and can also provide chaotic multiplexing capabilities, wherein two or more messages can modulate different circuit parameters and be sent and recovered using one transmission signal.
- Independent source modulation. This is another indirect modulation form where the information becomes an independent voltage/current source that is inserted in the chaotic transmitter circuit.
- Generalized modulation. This form involves the generalization of additive masking/multiplicative modulation, where the information and chaotic carrier are combined in a more general invertible manner.

Those modulation methods have been enhanced and extended to different applications in engineering. From the seminal work of Pecora and Carrol, two identical chaotic systems have been synchronized [77], and more recently, other synchronization schemes have been proposed to synchronize nonidentical and of different order chaotic systems. In this chapter, a generalized approach to synchronize chaotic systems is applied. That approach is based on the perspective of passivity-based state observer design in the context of generalized Hamiltonian systems including dissipation and destabilizing vector fields [80].

One of the early methods, called additive masking, used in constructing chaos-based secure communication systems, was based on simply adding the secret message to one of the chaotic states of the transmitter provided that the strength of the former is much weaker than that of the later [80]. This scheme is used herein to process data related to images in gray tones.

## 7.2 Hamiltonian Synchronization Approach

One hot topic for chaotic oscillators is their synchronization [81–83], which has received increased interest. This property is supposed to have interesting applications in different fields, particularly in designing secure communication systems [80, 84–86]. For instance, private communication schemes are usually composed of chaotic systems [80, 87–89], where the confidential information is embedded into the transmitted chaotic signal by direct modulation, masking, or another technique. At the receiver end, if chaotic synchronization is achieved, then it is possible to extract the hidden information from the transmitted signal.

Multi-scrolls chaotic attractors can be synchronized by applying Hamiltonian forms and observer approach. This technique is well described in the seminal article [80]. We adopt it because of its suitability to automation [88, 90].

Lets us consider the dynamical system

$$\dot{x} = f(x) \tag{7.1}$$

where $x \in \mathbb{R}^n$ is the state vector and $f : \mathbb{R}^n \to \mathbb{R}^n$ is a nonlinear function. In [80], it is reported that how does the system given in (7.1) can be written in the *Generalized Hamiltonian canonical form*:

$$\dot{x} = J(x)\frac{\partial H}{\partial x} + S(x)\frac{\partial H}{\partial x} + F(x), \quad x \in \mathbb{R}^n \tag{7.2}$$

where $H(x)$ denotes a smooth energy function which is globally positive definite in $\mathbb{R}^n$. The gradient vector of $H$, denoted by $\frac{\partial H}{\partial x}$, is assumed to exist everywhere. Quadratic energy function is used as $H(x) = \frac{1}{2}x^T Mx$ with $M$ being a constant, symmetric positive definite matrix. In this case, $\frac{\partial H}{\partial x} = Mx$. The matrices $J(x)$ and $S(x)$ satisfy, for all $x \in \mathbb{R}^n$, the following properties: $J(x) + J^T(x) = 0$ and $S(x) = S^T(x)$. The vector field $J(x)\frac{\partial H}{\partial x}$ exhibits the conservative part of the system and it is also referred to as the workless part, or workless forces of the system, and $J(x)$ denotes the working or nonconservative part of the system.

For certain systems, $S(x)$ is *negative definite* or *negative semidefinite*. Thus, the vector field is referred to as the dissipative part of the system. If, on the other hand, $S(x)$ is positive definite, positive semidefinite, or indefinite, it clearly represents the global, semi-global, or local destabilizing part of the system, respectively. In the last case, one can always (although nonuniquely) decompose such an indefinite symmetric matrix into the sum of a symmetric negative semidefinite matrix $R(x)$ and a symmetric positive semidefinite matrix $N(x)$. Finally, $F(x)$ represents a locally destabilizing vector field.

In the context of observer design, one can consider a special class of generalized Hamiltonian forms with output $y(t)$, given by

$$\begin{aligned} \dot{x} &= J(y)\frac{\partial H}{\partial x} + (I + S)\frac{\partial H}{\partial x} + F(y), \quad x \in \mathbb{R}^n \\ y &= C\frac{\partial H}{\partial x}, \quad y \in \mathbb{R}^m \end{aligned} \tag{7.3}$$

where $S$ is a constant symmetric matrix, not necessarily of a definite sign. $I$ is a constant skew symmetric matrix, and $C$ is a constant matrix.

The estimate of the state $x(t)$ can be denoted by $\xi(t)$, and one can consider the Hamiltonian energy function $H(\xi)$ to be the particularization of $H$ in terms of $\xi(t)$. Similarly, one can denote by $\eta(t)$ the estimated output, computed in terms of $\xi(t)$. The gradient vector $\frac{\partial H(\xi)}{\partial \xi}$ is, naturally, of the form $M\xi$ with $M$ being a constant, symmetric positive definite matrix.

A nonlinear state observer for the generalized Hamiltonian form (7.3) is given by

$$\begin{aligned} \dot{\xi} &= J(y)\frac{\partial H}{\partial \xi} + (I+S)\frac{\partial H}{\partial \xi} + F(y) + K(y-\eta), \quad \xi \in \mathbb{R}^n \\ \eta &= C\frac{\partial H}{\partial \xi}, \quad \eta \in \mathbb{R}^m \end{aligned} \tag{7.4}$$

where $K$ is the observer gain. The state estimation error, defined as $e(t) = x(t) - \xi(t)$, and the output estimation error, defined as $e_y(t) = y(t) - \eta(t)$, are governed by

$$\begin{aligned} \dot{e} &= J(y)\frac{\partial H}{\partial e} + (I+S-KC)\frac{\partial H}{\partial e}, \quad e \in \mathbb{R}^n \\ e_y &= C\frac{\partial H}{\partial e}, \quad e_y \in \mathbb{R}^m \end{aligned} \tag{7.5}$$

where $\frac{\partial H}{\partial e}$ actually stands, with some abuse of notation, for the gradient vector of the modified energy function, $\frac{\partial H(e)}{\partial e} = \frac{\partial H}{\partial x} - \frac{\partial H}{\partial \xi} = M(x-\xi) = Me$. We set, when needed, $I + S = W$.

**Definition 1** (*Chaotic synchronization*) [80] The slave system (nonlinear state observer) (7.4) synchronizes with the chaotic master system in generalized Hamiltonian form (7.3), if

$$\lim_{t\to\infty} \|x(t) - \xi(t)\| = 0 \tag{7.6}$$

no matter which initial conditions $x(0)$ and $\xi(0)$ have, where the state estimation error $e(t) = x(t) - \xi(t)$ corresponds to the synchronization error.

## 7.3 Synchronization of Multi-scroll Chaotic Attractors

The chaos generator model (7.1)–(7.3) in generalized Hamiltonian form is given by

$$\begin{bmatrix} \dot{x}_1 \\ \dot{x}_2 \\ \dot{x}_3 \end{bmatrix} = \begin{bmatrix} 0 & \frac{1}{2b} & \frac{1}{2} \\ -\frac{1}{2b} & 0 & 1 \\ -\frac{1}{2} & -1 & 0 \end{bmatrix} \frac{\partial H}{\partial x} + \begin{bmatrix} 0 & \frac{1}{2b} & -\frac{1}{2} \\ \frac{1}{2b} & 0 & 0 \\ -\frac{1}{2} & 0 & -c \end{bmatrix} \frac{\partial H}{\partial x} + \begin{bmatrix} 0 \\ 0 \\ d_1 f(x) \end{bmatrix} \tag{7.7}$$

One can take as Hamiltonian energy function

$$H(x) = \frac{1}{2}\left[ax_1^2 + bx_2^2 + x_3^2\right] \tag{7.8}$$

and as gradient vector

$$\frac{\partial H}{\partial x} = \begin{bmatrix} a & 0 & 0 \\ 0 & b & 0 \\ 0 & 0 & 1 \end{bmatrix} \begin{bmatrix} x_1 \\ x_2 \\ x_3 \end{bmatrix} = \begin{bmatrix} ax_1 \\ bx_2 \\ x_3 \end{bmatrix}$$

The destabilizing vector field calls for $x_1$ and $x_2$ signals to be used as the outputs, of the master model (7.7). One can use $y = x_1$ in (7.7). The matrices $C$, $S$, and $I$ are given by

$$C = \begin{bmatrix} \frac{1}{a} & 0 & 0 \end{bmatrix}$$
$$S = \begin{bmatrix} 0 & \frac{1}{2b} & -\frac{1}{2} \\ \frac{1}{2b} & 0 & 0 \\ -\frac{1}{2} & 0 & -c \end{bmatrix}$$
$$I = \begin{bmatrix} 0 & \frac{1}{2b} & \frac{1}{2} \\ -\frac{1}{2b} & 0 & 1 \\ -\frac{1}{2} & -1 & 0 \end{bmatrix}$$

The pair $(C, S)$ is observable. Therefore, the nonlinear state observer for (7.7), to be used as the slave model, is designed according to (7.4) as

$$\begin{bmatrix} \dot{\xi}_1 \\ \dot{\xi}_2 \\ \dot{\xi}_3 \end{bmatrix} = \begin{bmatrix} 0 & \frac{1}{2b} & \frac{1}{2} \\ -\frac{1}{2b} & 0 & 1 \\ -\frac{1}{2} & -1 & 0 \end{bmatrix} \frac{\partial H}{\partial \xi} + \begin{bmatrix} 0 & \frac{1}{2b} & -\frac{1}{2} \\ \frac{1}{2b} & 0 & 0 \\ -\frac{1}{2} & 0 & -c \end{bmatrix} \frac{\partial H}{\partial \xi} + \dots + \begin{bmatrix} 0 \\ 0 \\ d_1 f(\xi) \end{bmatrix} + \begin{bmatrix} k_1 \\ k_2 \\ k_3 \end{bmatrix} e_y \tag{7.9}$$

with gain $k_i$, $i = 1, 2, 3$ to be selected in order to guarantee asymptotic exponential stability to zero of the state reconstruction error trajectories (i.e., synchronization error $e(t)$). From (7.7) and (7.9) one have that the synchronization error dynamics is governed by [88]

$$\begin{bmatrix} \dot{e}_1 \\ \dot{e}_2 \\ \dot{e}_3 \end{bmatrix} = \begin{bmatrix} 0 & \frac{1}{2b} & \frac{1}{2} \\ -\frac{1}{2b} & 0 & 1 \\ -\frac{1}{2} & -1 & 0 \end{bmatrix} \frac{\partial H}{\partial e} + \begin{bmatrix} 0 & \frac{1}{2b} & -\frac{1}{2} \\ \frac{1}{2b} & 0 & 0 \\ -\frac{1}{2} & 0 & -c \end{bmatrix} \frac{\partial H}{\partial e} + \begin{bmatrix} k_1 \\ k_2 \\ k_3 \end{bmatrix} e_y \tag{7.10}$$

As an example: By selecting $K = (k_1, k_2, k_3)^T$ with $k_1 = 2$, $k_2 = 5$, $k_3 = 7$, and considering the initial condition $X(0) = [0, 0, 0.1]$, $\xi(0) = [1, -0.5, 3]$, then one can carry out numerical simulations using a numerical method like $ode45$ in MATLAB with a full integration of $T = 2000$ to generate a 4-scroll chaotic oscillator. Figures 7.1a, b and 7.2a, b show the state trajectories between the master and slave models (7.7) and (7.9), with different maximum Lyapunov exponents (MLE), respec-

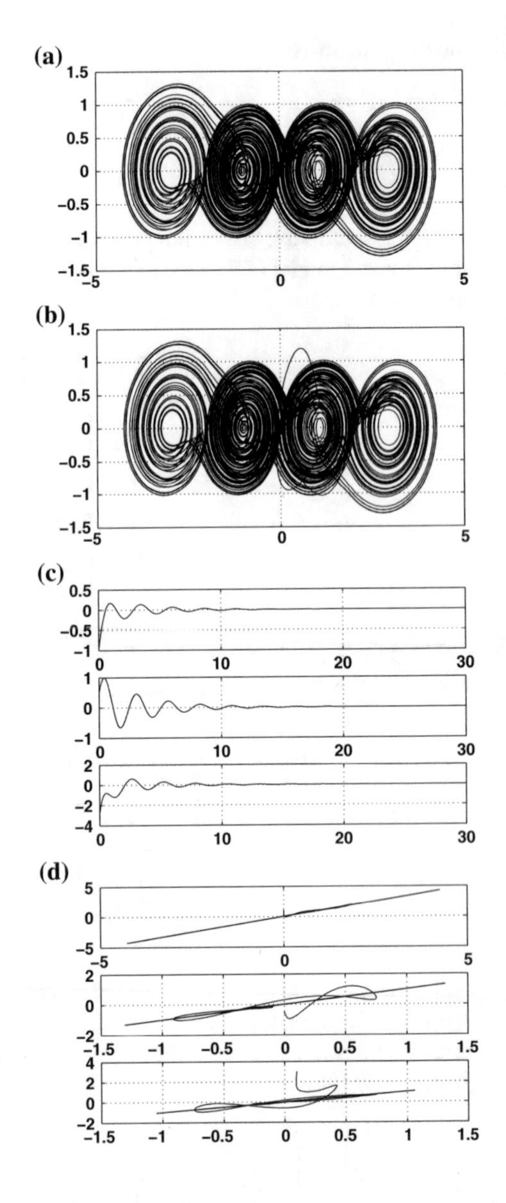


**Fig. 7.1** 4-scroll chaotic oscillators master–slave synchronization with coefficients equal to 0.7. **a** Master. **b** Slave. **c** Error synchronization transient evolution. **d** Error phase diagram of the synchronized states $x$ and $\xi$

tively. Figures 7.1c, d and 7.2c, d show the synchronization error that is modeled by (7.10), and the phase error between the master and slave chaotic oscillators, respectively. As one sees, when MLE is low, the synchronization is accomplished earlier than when MLE is higher. The former case needs around 10 iterations to synchronize, while when MLE is high more than 20 iterations are required to synchronize. This is a trade-off in designing multi-scroll chaotic oscillators, but a high MLE ensures better unpredictability.

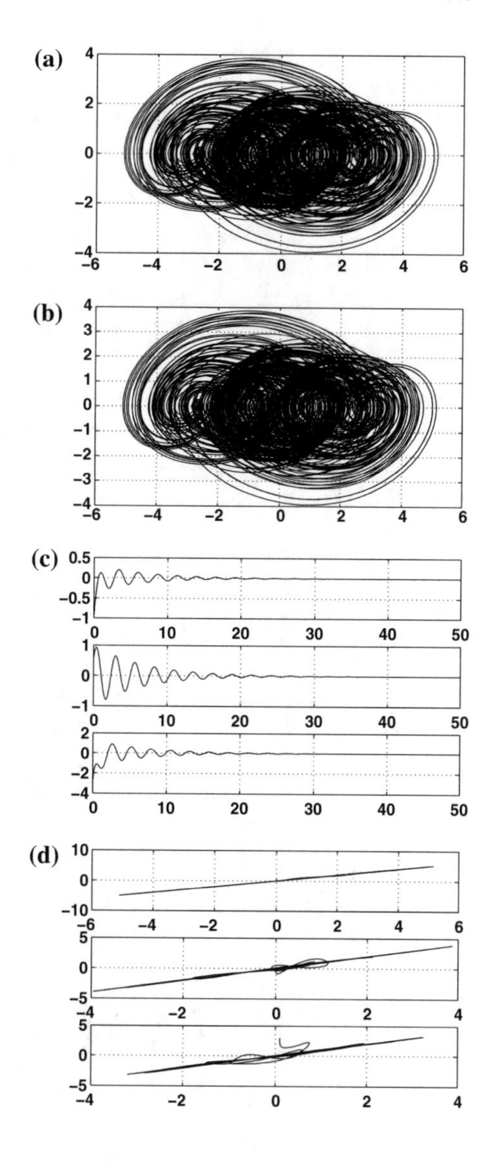


**Fig. 7.2** Optimized MLE for the 4-scroll chaotic oscillator master–slave synchronization. **a** Master. **b** Slave. **c** Error synchronization transient evolution. **d** Error phase diagram of the synchronized states $x$ and $\xi$

## 7.4 Synchronization of 2D-4-Scroll Chaos Generators

The synchronization of PWL-based chaotic oscillators can be extended to more directions, as the cases already detailed in [91], where the 2-direction (2D) multi-scroll chaotic system defined by (7.11), is used to perform master–slave synchronization by Hamiltonian forms and observer approach [80]. First, one can propose a Hamilton energy function and its gradient vector as given in (7.12) and (7.13), respectively.

$$\begin{aligned}\dot{x} &= y - \tfrac{d_2}{b} f(y; k_2, h_2, p_2, q_2)\\ \dot{y} &= z\\ \dot{z} &= -ax - by - cz + d_1 f(x; k_1, h_1, p_1, q_1) + d_2 f(y; k_2, h_2, p_2, q_2),\end{aligned} \tag{7.11}$$

$$H(x) = \frac{1}{2}\left[ax_1^2 + bx_2^2 + x_3^2\right], \tag{7.12}$$

$$\frac{\partial H}{\partial x} = \begin{bmatrix} ax_1 \\ bx_2 \\ x_3 \end{bmatrix}, \tag{7.13}$$

Second, one can obtain the matrices $S$ and $J$ as shown in (7.14) and (7.15), respectively. That way, the 2D chaos generator in (7.11) can be described in generalized Hamiltonian forms, as shown in (7.16) and (7.17), where (7.17) is the nonlinear state observer of (7.16).

$$S(x) = \frac{1}{2}\left\{\begin{bmatrix} 0 & \frac{1}{b} & 0 \\ 0 & 0 & 1 \\ -1 & -1 & -c \end{bmatrix} + \begin{bmatrix} 0 & 0 & -1 \\ \frac{1}{b} & 0 & -1 \\ 0 & 1 & -c \end{bmatrix}\right\} = \begin{bmatrix} 0 & \frac{1}{2b} & -\frac{1}{2} \\ \frac{1}{2b} & 0 & 0 \\ -\frac{1}{2} & 0 & -c \end{bmatrix} \tag{7.14}$$

$$J(x) = \frac{1}{2}\left\{\begin{bmatrix} 0 & \frac{1}{b} & 0 \\ 0 & 0 & 1 \\ -1 & -1 & -c \end{bmatrix} - \begin{bmatrix} 0 & 0 & -1 \\ \frac{1}{b} & 0 & -1 \\ 0 & 1 & -c \end{bmatrix}\right\} = \begin{bmatrix} 0 & \frac{1}{2b} & \frac{1}{2} \\ -\frac{1}{2b} & 0 & 1 \\ -\frac{1}{2} & -1 & 0 \end{bmatrix}. \tag{7.15}$$

$$\begin{bmatrix} \dot{x}_1 \\ \dot{x}_2 \\ \dot{x}_3 \end{bmatrix} = \begin{bmatrix} 0 & \frac{1}{2b} & \frac{1}{2} \\ -\frac{1}{2b} & 0 & 1 \\ -\frac{1}{2} & -1 & 0 \end{bmatrix}\frac{\partial H}{\partial x} + \begin{bmatrix} 0 & \frac{1}{2b} & -\frac{1}{2} \\ \frac{1}{2b} & 0 & 0 \\ -\frac{1}{2} & 0 & -c \end{bmatrix}\frac{\partial H}{\partial x} + \begin{bmatrix} -\frac{d_2}{b} f(x_2) \\ 0 \\ d_1 f(x_1) + d_2 f(x_2) \end{bmatrix} \tag{7.16}$$

$$\begin{aligned}\begin{bmatrix} \dot{\xi}_1 \\ \dot{\xi}_2 \\ \dot{\xi}_3 \end{bmatrix} &= \begin{bmatrix} 0 & \frac{1}{2b} & \frac{1}{2} \\ -\frac{1}{2b} & 0 & 1 \\ -\frac{1}{2} & -1 & 0 \end{bmatrix}\frac{\partial H}{\partial \xi} + \begin{bmatrix} 0 & \frac{1}{2b} & -\frac{1}{2} \\ \frac{1}{2b} & 0 & 0 \\ -\frac{1}{2} & 0 & -c \end{bmatrix}\frac{\partial H}{\partial \xi} \\ &+ \begin{bmatrix} -\frac{d_2}{b} f(x_2) \\ 0 \\ d_1 f(x_1) + d_2 f(x_2) \end{bmatrix} + \begin{bmatrix} k_1 & k_4 \\ k_2 & k_5 \\ k_3 & k_6 \end{bmatrix}(y - \eta).\end{aligned} \tag{7.17}$$

with $\eta$ being

$$\eta = \begin{bmatrix} \frac{d_1}{a} & \frac{d_2}{b} & 0 \\ 0 & \frac{d_2}{b^2} & 0 \end{bmatrix}\frac{\partial H}{\partial \xi} \tag{7.18}$$

By evaluating the stability of the approach using Observability's criterion, one obtains

$$det \begin{vmatrix} \frac{d_1}{a} & \frac{d_2}{2b^2} & \frac{\left(\frac{1}{4b^2}+\frac{1}{4b}\right)d_1}{a} \\ \frac{d_2}{b} & \frac{d_1}{2ab} & \frac{d_2}{4b^3} \\ 0 & -\frac{d_1}{2a} & \frac{cd_1}{2a}-\frac{d_2}{4b^2} \end{vmatrix} \neq 0, \tag{7.19}$$

Besides, it is also necessary to demonstrate Theorem 7.2, in order to gain insight about the synchronization of 2D chaos generators. So that matrices $S$, $C$, and $K$ are used to evaluate the equation in Theorem 7.2, resulting on (7.22).

**Theorem 7.1** *The state $x$ of the nonlinear system (7.20) can be globally, exponentially, asymptotically estimated by the state $\xi$ of an observer of the form (7.21), if the pair of matrices $C$, $S$ are either observable or, at least, detectable.*

**Theorem 7.2** *The state $x$ of the nonlinear system (7.20) can be globally, exponentially, asymptotically estimated, by the state $\xi$ of the observer (7.21) if and only if there is a constant matrix $K$; such as $[W - KC] + [W - KC]^T = 2[S - \frac{1}{2}(KC - C^T K^T)]$, and is negative definite.*

$$\begin{aligned} \dot{x} &= J(y)\tfrac{\partial H}{\partial x} + S(y)\tfrac{\partial H}{\partial x} + F(y), \quad x \in R^n \\ y &= C\tfrac{\partial H}{\partial x}, \quad y \in R^m \end{aligned} \tag{7.20}$$

In (7.20) $S$ is a constant and symmetric matrix, $y$ the output vector of the system and $C$ a constant matrix. By selecting $\xi$ as the estimated state vector of $x$, and $\eta$ as the estimated output in terms of $\xi$; a dynamic nonlinear state observer for (7.20) is given in (7.21); with $K$ being a constant vector, known as the observer gain.

$$\begin{aligned} \dot{\xi} &= J(y)\tfrac{\partial H}{\partial \xi} + S(y)\tfrac{\partial H}{\partial \xi} + F(y) + K(y - \eta), \\ \eta &= C\tfrac{\partial H}{\partial \xi}. \end{aligned} \tag{7.21}$$

$$2\left[\begin{bmatrix} 0 & \frac{1}{2b} & -\frac{1}{2} \\ \frac{1}{2b} & 0 & 0 \\ -\frac{1}{2} & 0 & -c \end{bmatrix} - \frac{1}{2}\left\{\begin{bmatrix} k_1 & k_4 \\ k_2 & k_5 \\ k_3 & k_6 \end{bmatrix}\begin{bmatrix} \frac{d_1}{a} & \frac{d_2}{b} & 0 \\ 0 & \frac{d_2}{b^2} & 0 \end{bmatrix} + \begin{bmatrix} \frac{d_1}{a} & 0 \\ \frac{d_2}{b} & \frac{d_2}{b^2} \\ 0 & 0 \end{bmatrix}\begin{bmatrix} k_1 & k_2 & k_3 \\ k_4 & k_5 & k_6 \end{bmatrix}\right\}\right]$$
$$= \begin{bmatrix} -\frac{2k_1d_1}{a} & \frac{1}{b}-\frac{k_1d_2}{b}-\frac{k_4d_2}{b^2}-\frac{k_2d_1}{a} & -1-\frac{k_3d_1}{a} \\ \frac{1}{b}-\frac{k_1d_2}{b}-\frac{k_4d_2}{b^2}-\frac{k_2d_1}{a} & -\frac{2k_2d_2}{b}-\frac{k_5d_2}{b^2} & -\frac{k_3d_2}{b}-\frac{k_6d_2}{b^2} \\ -\frac{1}{b}-\frac{k_3d_1}{a} & -\frac{k_3d_2}{b}-\frac{k_6d_2}{b^2} & -2c \end{bmatrix} \tag{7.22}$$

The Sylvester's criterion is used to demonstrate that the matrix in (7.22) is negative definite. Indeed, the values for the observer gain, matrix in (7.17) are also obtained by calculating the roots of the determinants in (7.22). For the first determinant, one obtains

$$-\frac{2k_1d_1}{a} < 0 \Longrightarrow k_1 > 0 \tag{7.23}$$

Equation (7.24) is obtained by solving for the minor of the matrix $2 \times 2$ in (7.22).

$$\begin{aligned} det = &-\frac{1}{b^4a^2}\left(-2k_1d_1d_2b^3ak_2 - 4k_1d_1d_2b^2ak_5 + b^2a^2 - 2b^2a^2k_1d_2 - 2ba^2k_4d_2\right. \\ &\left.-2b^3ak_2d_1 + k_1^2d_2^2b^2a^2 + 2k_1d_2^2ba^2k_4 + k_4^2d_2^2a^2 + 2k_4d_2ak_2d_1b^2 + k_2^2d_1^2b^4\right) < 0 \end{aligned} \tag{7.24}$$

Considering that $k_1 = k_4$, $k_2 = k_5$, and $k_3 = k_6$, (7.24) can be updated by (7.25). This assumption is valid since the nonlinear functions $f(x_1)$ and $f(x_2)$ in (7.15) are the same.

$$\begin{aligned} det2 = &-\frac{1}{b^4a^2}\left(-2k_1d_1d_2b^3ak_2 - 2k_1d_1d_2b^2ak_2 + b^2a^2 - 2b^2a^2k_1d_2 - 2ba^2k_1d_2\right. \\ &\left.-2b^3ak_2d_1 + k_1^2d_2^2b^2a^2 + 2k_1^2d_2^2ba^2 + k_1^2d_2^2a^2 + k_2^2d_1^2b^4\right) < 0 \end{aligned} \tag{7.25}$$

By solving (7.25), the interval values for $k_2$ are obtained as shown in (7.26).

$$\begin{aligned} &\frac{\left(b + k_1d_2b + k_1d_2 - 2\sqrt{b^2k_1d_2 + k_1d_2b}\right)a}{d_1b^2} < k_2 \\ &< \frac{\left(b + k_1d_2b + k_1d_2 + 2\sqrt{b^2k_1d_2 + k_1d_2b}\right)a}{d_1b^2}. \end{aligned} \tag{7.26}$$

Note that $k_3$ has no influence on the observable eigenvalues of the nonconservative structure of the 2D chaos generator.

## 7.5 Synchronization of 3D-4-Scroll Chaos Generators

Lets us consider the 3D-multi-scroll chaotic system defined by (7.27), with $f(x_1)$, $f(x_2)$, and $f(x_3)$ being saturated nonlinear function series approached by PWL functions.

$$\begin{aligned} \dot{x}_1 &= x_2 - \tfrac{d_2}{b}f(x_2) \\ \dot{x}_2 &= x_3 - \tfrac{d_3}{b}f(x_3) \\ \dot{x}_3 &= -ax_1 - bx_2 - cx_3 + d_1f(x_1) + d_2f(x_2) + d_3f(x_3), \end{aligned} \tag{7.27}$$

where $x_1, x_2, x_3$ are state variables, and $a, b, c, d_1, d_2, d_3 = 0.7$ are positive real constants. Using (7.20), (7.21), and (7.13), the 3D chaos generator in (7.27) can be described in generalized Hamiltonian forms as shown in (7.28). Consequently, the nonlinear state observer for 3D chaos generator in (7.27), according to (7.21), is shown in (7.29).

$$\begin{bmatrix} \dot{x}_1 \\ \dot{x}_2 \\ \dot{x}_3 \end{bmatrix} = \begin{bmatrix} 0 & \frac{1}{2b} & \frac{1}{2} \\ -\frac{1}{2b} & 0 & 1 \\ -\frac{1}{2} & -1 & 0 \end{bmatrix} \frac{\partial H}{\partial x} + \begin{bmatrix} 0 & \frac{1}{2b} & -\frac{1}{2} \\ \frac{1}{2b} & 0 & 0 \\ -\frac{1}{2} & 0 & -c \end{bmatrix} \frac{\partial H}{\partial x}$$

$$+ \begin{bmatrix} -\frac{d_2}{b} f(x_2) \\ -\frac{d_3}{b} f(x_3) \\ d_1 f(x_1) + d_2 f(x_2) + d_3 f(x_3) \end{bmatrix} \tag{7.28}$$

$$\begin{bmatrix} \dot{\xi}_1 \\ \dot{\xi}_2 \\ \dot{\xi}_3 \end{bmatrix} = \begin{bmatrix} 0 & \frac{1}{2b} & \frac{1}{2} \\ -\frac{1}{2b} & 0 & 1 \\ -\frac{1}{2} & -1 & 0 \end{bmatrix} \frac{\partial H}{\partial \xi} + \begin{bmatrix} 0 & \frac{1}{2b} & -\frac{1}{2} \\ \frac{1}{2b} & 0 & 0 \\ -\frac{1}{2} & 0 & -c \end{bmatrix} \frac{\partial H}{\partial \xi}$$

$$+ \begin{bmatrix} -\frac{d_2}{b} f(x_2) \\ -\frac{d_3}{b} f(x_3) \\ d_1 f(x_1) + d_2 f(x_2) + d_3 f(x_3) \end{bmatrix} + \begin{bmatrix} k_1 & k_4 & k_7 \\ k_2 & k_5 & k_8 \\ k_3 & k_6 & k_9 \end{bmatrix} (y - \eta). \tag{7.29}$$

with $\eta$ being

$$\eta = \begin{bmatrix} \frac{d_1}{a} & \frac{d_2}{b} & \frac{d_3}{b} \\ 0 & \frac{d_2}{b^2} & 0 \\ 0 & 0 & \frac{d_3}{b} \end{bmatrix} \frac{\partial H}{\partial \xi} \tag{7.30}$$

According to (7.19), the approach is stable and the demonstration of Theorem 7.2 for the synchronization of three-directional chaos generators is given in (7.31), using the matrices $S$, $K$, and $C$ in (7.14), (7.29), and (7.30), respectively.

$$\begin{bmatrix} -\frac{2k_1d_1}{a} & \frac{1}{b} - \frac{k_1d_2}{b} - \frac{k_4d_2}{b^2} - \frac{k_2d_1}{a} & -1 - \frac{k_3d_1}{a} - \frac{k_1d_3}{b} - \frac{k_7d_3}{b} \\ \frac{1}{b} - \frac{k_1d_2}{b} - \frac{k_4d_2}{b^2} - \frac{k_2d_1}{a} & -\frac{2k_2d_2}{b} - \frac{2k_5d_2}{b^2} & -\frac{k_2d_3}{b} - \frac{k_8d_3}{b} - \frac{k_3d_2}{b} - \frac{k_6d_2}{b^2} \\ -\frac{1}{b} - \frac{k_3d_1}{a} - \frac{k_1d_3}{b} - \frac{k_7d_3}{b} & -\frac{k_2d_3}{b} - \frac{k_8d_3}{b} - \frac{k_3d_2}{b} - \frac{k_6d_2}{b^2} & -2c - \frac{2k_3d_3}{b} - \frac{k_9d_3}{b} \end{bmatrix} \tag{7.31}$$

Similarly, the values of the observer gain are calculated using the Sylvester's criterion. Note that the first two determinants in (7.31) are identical to that on (7.22), since the extra nonlinear function $f(x_3)$ is only related to state variable $x_3$. In this manner, the intervals for $k_1$ and $k_2$ are also given by (7.23) and (7.25), respectively. Now, consider that $k_1 = k_4 = k_7$, $k_2 = k_5 = k_8$, and $k_3 = k_6 = k_9$, since the nonlinear functions

$f(x_1)$, $f(x_2)$, and $f(x_3)$ in (7.29) are the same. By evaluating (7.29), it is found the interval of values for $k_3$ shown in (7.33).

$$det_{3\times3} = 16.58k_3 - 9.71k_1 + 9.78k_2 + 0.18k_2k_3 - 3.45k_2^2 + 6.93k_3^2 + 16.42k_1k_2$$
$$-(2e-9)k_1k_3^2 - 28.20k_3k_1 + 8.25k_1^2 + (2e-9)k_1k_2k_3 + 2.85 < 0 \tag{7.32}$$

$$k_3 = \frac{4.14e9}{-3.46e9 + k_1} + \frac{4.59e7k_2}{-3.46e9 + k_1} - \frac{4.05e9k_1}{-3.46e9 + k_1} + \frac{0.50k_1k_2}{-3.46e9 + k_1}$$
$$+ \frac{1}{-3.46e9 + k_1}\big(0.50\big(4.89e19 - 6.7e19k_2 - 1.66e20k_1 - 1.16e20k_1k_2$$
$$+ 2.39e19k_2^2 - 6.73e9k_1k_2^2 + 1.41e20k_1^2 4.64e9k_1^2k_2 + k_1^2k_2^2 + 1.65e10k_1^3\big)^{1/2}\big) \tag{7.33}$$

### 7.5.1 Numerical Simulation Results

By selecting $k_1 = k_4 = 1$, $k_2 = k_5 = 2$, and $k_3 = k_6 = 0$ for the observer in (7.17), one obtains the synchronization of 2D-4-scroll chaos generators, the coincidence of their states is represented by a straight line, with slope equal to unity, in the phase plane for each state and its error $e_x$, which is the difference between the observed state $x$ and the estimated state $\xi$, as shown in Figs. 7.3 and 7.4, respectively.

By selecting $k_1 = k_4 = k_7 = 1$, $k_2 = k_5 = k_8 = 2$, and $k_3 = k_6 = k_9 = 0$ for the observer in (7.29); one obtains Figs. 7.5 and 7.6.

A prediction of the values for the observer gain is shown in Fig. 7.7. The solution is calculated from (7.26) and any value selected between solution (+) and solution (−) as shown in Fig. 7.7 leads to the synchronization for 2D- and 3D-multi-scroll chaos generators given in (7.11) and (7.27).

## 7.6 Image Transmission Through a Chaotic Secure Communication System

The FPGA realization of chaos generators by applying different numerical methods has already been introduced in [25]. Further, the application to secure communications is given in [33], where two multi-scroll chaotic oscillators generating 2 and 6 scrolls are synchronized by applying Hamiltonian forms and observer approach. The

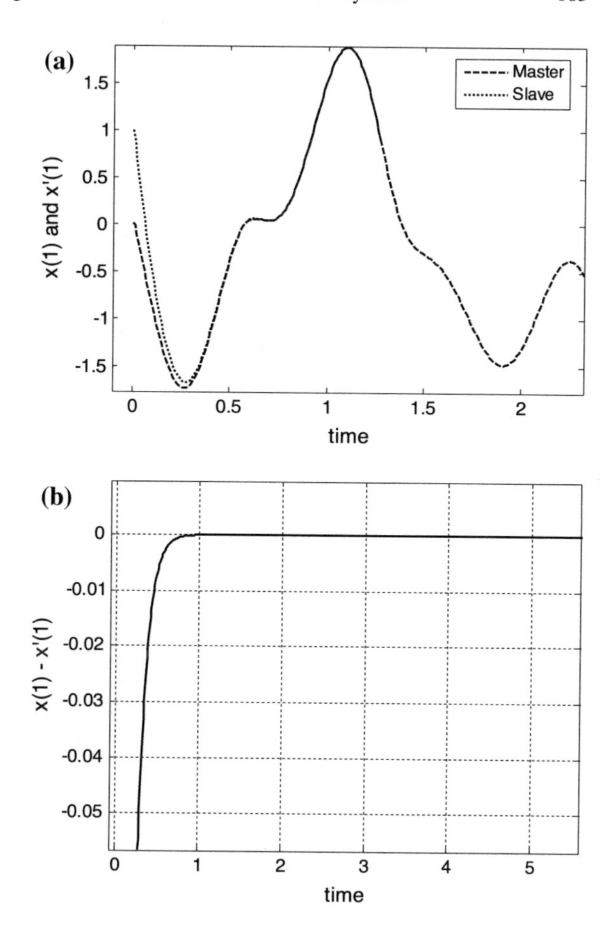


**Fig. 7.3** **a** Observed and estimated states in time domain for two 2D-4-scroll chaos generators. **b** Error between the synchronized 2D-4-scroll chaos generators

synchronized master–slave topology is used to implement a secure communication system by adding chaos to an image at the transmission stage and by subtracting chaos at the recover stage. This scheme is reproduced herein and three kinds of multi-scroll chaotic oscillators are shown before showing transmission of images in black and white and gray tones.

### 7.6.1 Multi-scroll Chaos Generators Based on PWL Functions

In Chap. 4, three PWL functions were detailed to generate chaotic attractors, they are: saturated nonlinear function series, sawtooth, and Chua's diode. The last one is based on PWL functions with negative slopes. They can be used in third-order continuous-time dynamical systems to generate chaos, and numerical methods solve

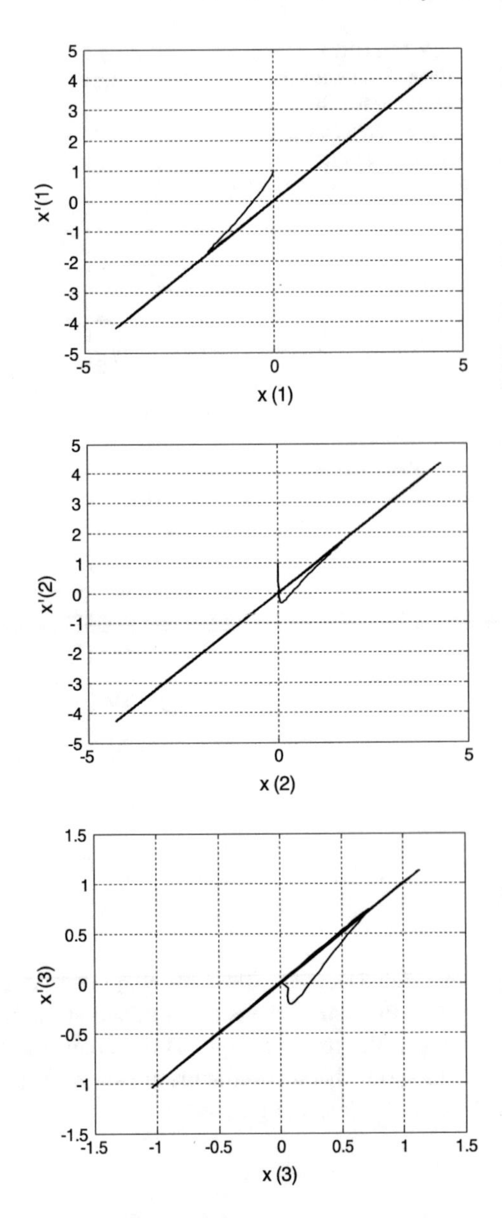


**Fig. 7.4** Phase plane diagrams for the state variables in (7.16) and (7.17)

those mathematical models. For instance, Forward Euler and Runge–Kutta methods have been applied in [25] to solve nonlinear dynamical systems. Fourth order Runge–Kutta requires evaluating four variables, which are described by the iterative formulae in (7.34) and (7.35), where one can appreciate operations like addition, subtraction, multiplication, and division, which are costly when implementing in

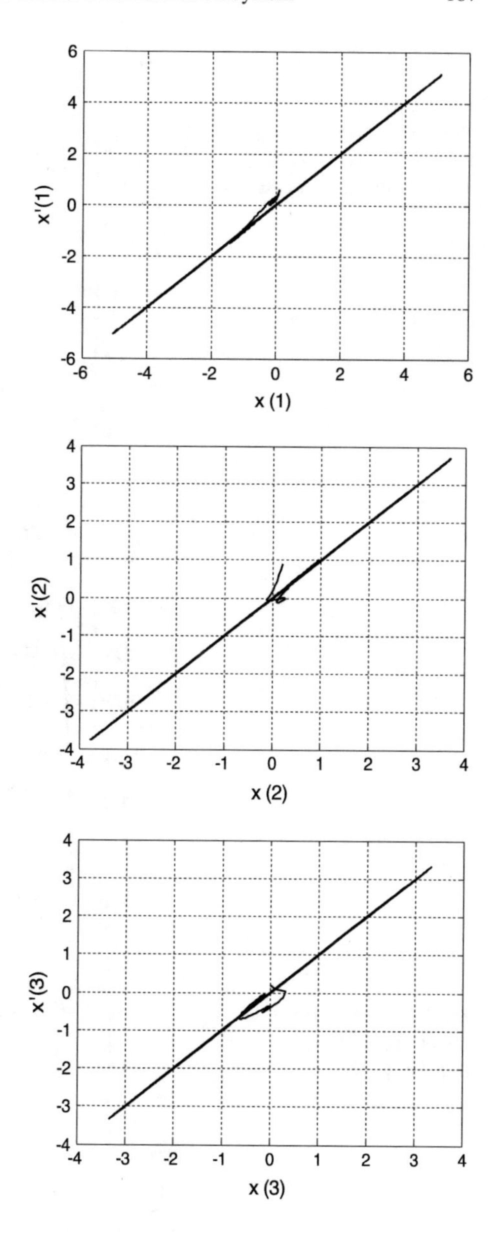


**Fig. 7.5** Phase plane diagrams for the state variables in (7.28) and (7.29)

FPGA, so that a designer should look for the best hardware realization to minimize hardware resources.

$$y_{n+1} = y_n + \frac{1}{6}h(k_1 + k_2 + k_3 + k_4) \tag{7.34}$$

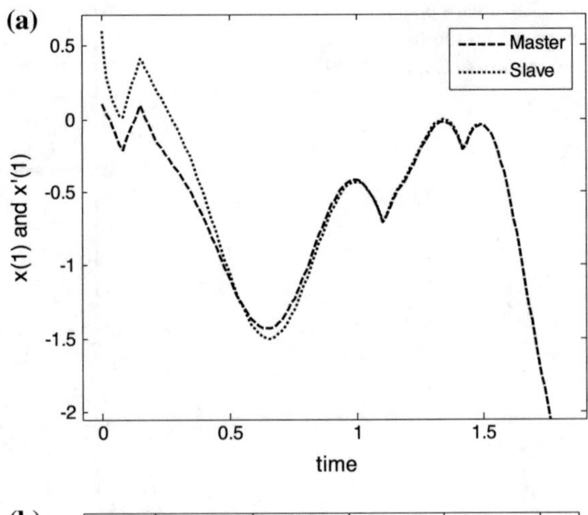


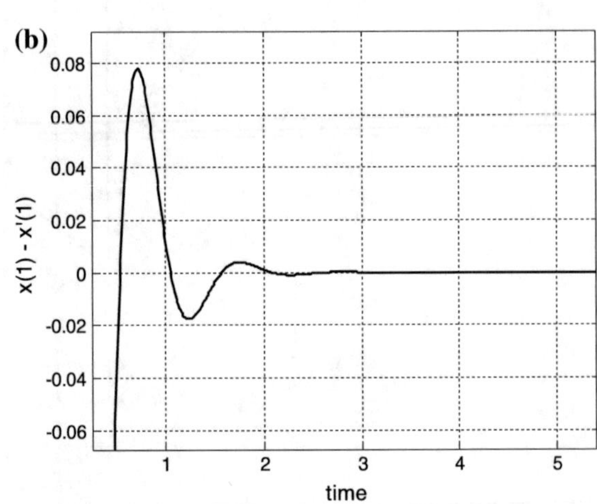


**Fig. 7.6** **a** Observed and estimated states in time domain for two 3D-4-scroll chaos generators. **b** Error between the synchronized 3D-4-scroll chaos generators

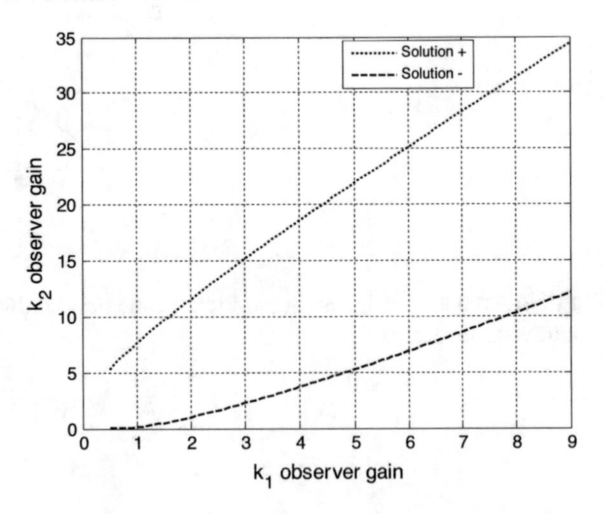


**Fig. 7.7** Synchronization region according to the observer gain

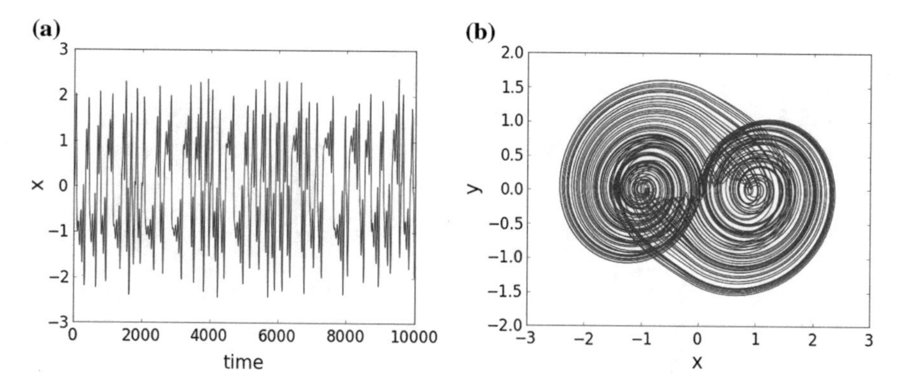


**Fig. 7.8** Saturated nonlinear functions series as PWL function to generate 2-scrolls. **a** State variable $x$. **b** Portrait $x$–$y$

$$
\begin{aligned}
k_1 &= f(x_n, y_n)\\
k_2 &= f(x_n, \frac{1}{2}h, y_n + \frac{1}{2}hk_1)\\
k_3 &= f(x_n, \frac{1}{2}h, y_n + \frac{1}{2}hk_2)\\
k_4 &= f(x_n, \frac{1}{2}h, y_n + \frac{1}{2}hk_3)
\end{aligned}
\tag{7.35}
$$

Figure 7.8 shows the state variable $x$ of the chaos generator using saturated series as PWL function. Figure 7.9 shows the state variable $x$ for the chaos generator using Chua's diode as PWL function. Figure 7.10 shows the state variable $x$ of the chaos generator using sawtooth function. In all cases the phase space portrait is shown between the state variables $x$–$y$. As one sees, all 2-scroll attractors can be used to

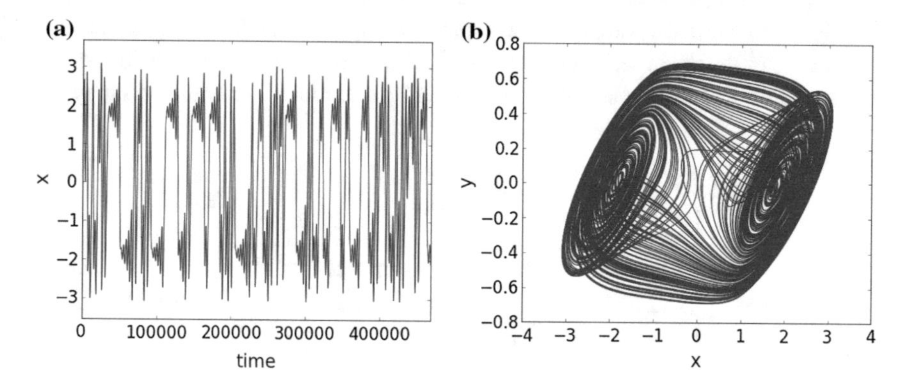


**Fig. 7.9** Chua's diode as PWL function to generate 2-scrolls. **a** State variable $x$. **b** Portrait $x$–$y$

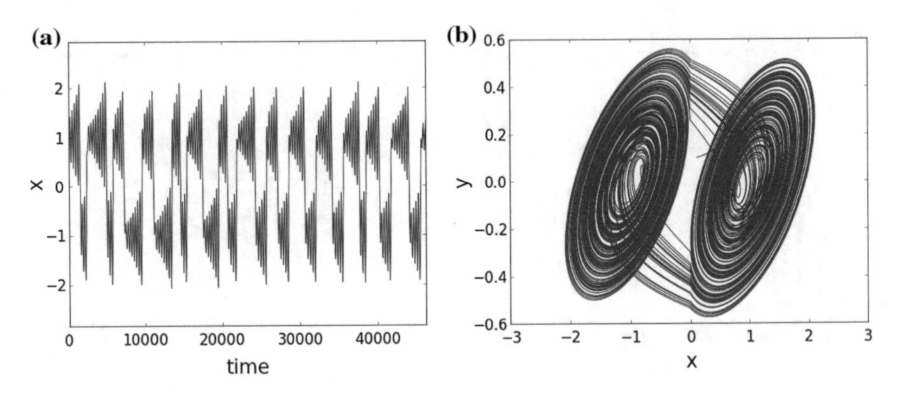


**Fig. 7.10** Sawtooth as PWL function to generate 2-scrolls. **a** Seal $x$. **b** Attractor $x$–$y$

implement a secure communication system, where the security should depend on the unpredictability of the chaos generators that is evaluated by MLE.

## 7.6.2 FPGA Realization

As already shown in [33, 25], the FPGA realization depends on the numerical method that is used to solve the system of equations. In any case, the basic building blocks are related to deal with adders, subtractors, multipliers, shift registers, multiplexers, and so on. Again, the three basic blocks are shown in Fig. 7.11, which include pins for clock (CLK) and reset (RST). The computer arithmetic was established to fixed point of 28-bits in a format 4.24.

According to the discretized equations, the VHDL blocks are interconnected. One important step is counting the number of blocks in series connection to estimate clock cycles since all blocks are synchronous ones. In this implementation, a multiplexer is designed to include a counter register that is waiting the number of pulses equal to the number of blocks in series connection. That way, the correct output is the correct one and the architecture can follow with the next iteration. The high-level diagram

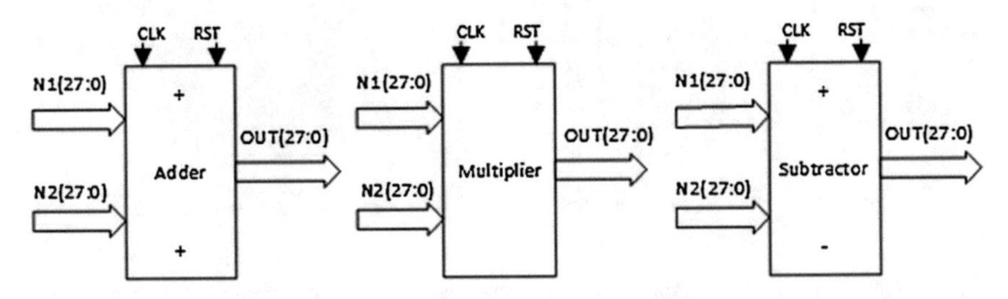


**Fig. 7.11** VHDL blocks

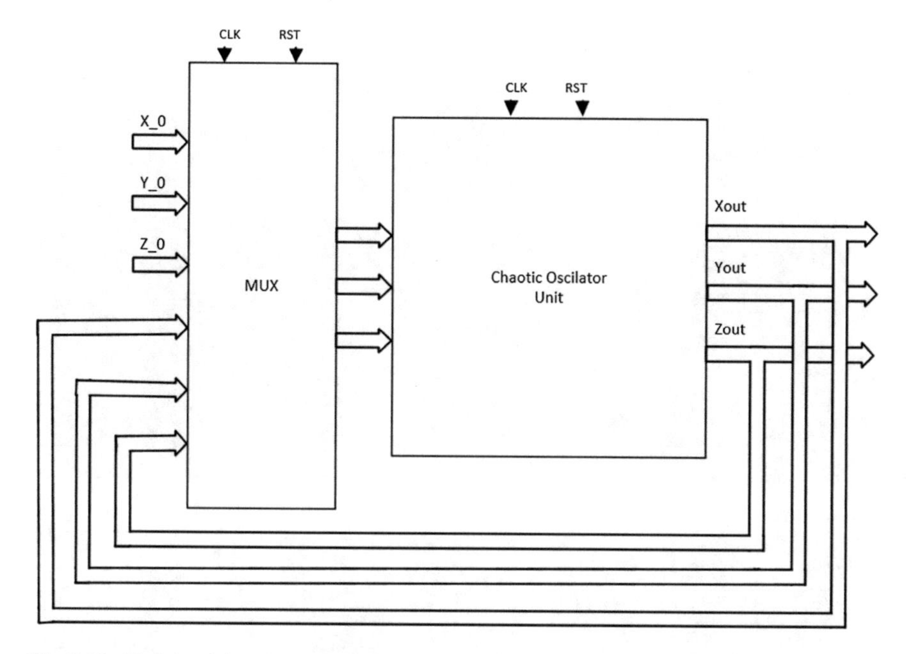


**Fig. 7.12** High-level description for the chaos generator

is shown in Fig. 7.12, where the multiplexer is used to set the initial conditions for solving the mathematical model of the chaos generator. As one can infer, once the initial conditions are used, the multiplexer connects the loop for performing the rest of iterations to generate chaotic behavior.

The experimental results for generating 2-scroll attractors from the FPGA realizations are shown in Figs. 7.13, 7.14, and 7.15.

### *7.6.3 Master–Slave Synchronization*

Taking as case of study the chaos generator based on saturated nonlinear function series as PWL function, this subsection shows the synchronization by Hamiltonian forms and observer approach. As shown above, the synchronization requires the formulation of the system of equations for the master and slave, as shown by (7.36) for the master oscillator, and (7.37) for the slave. Those equations are solved herein by applying the Forward Euler method, so that they include the step size $h$ and the gains of the observers $k_i$.

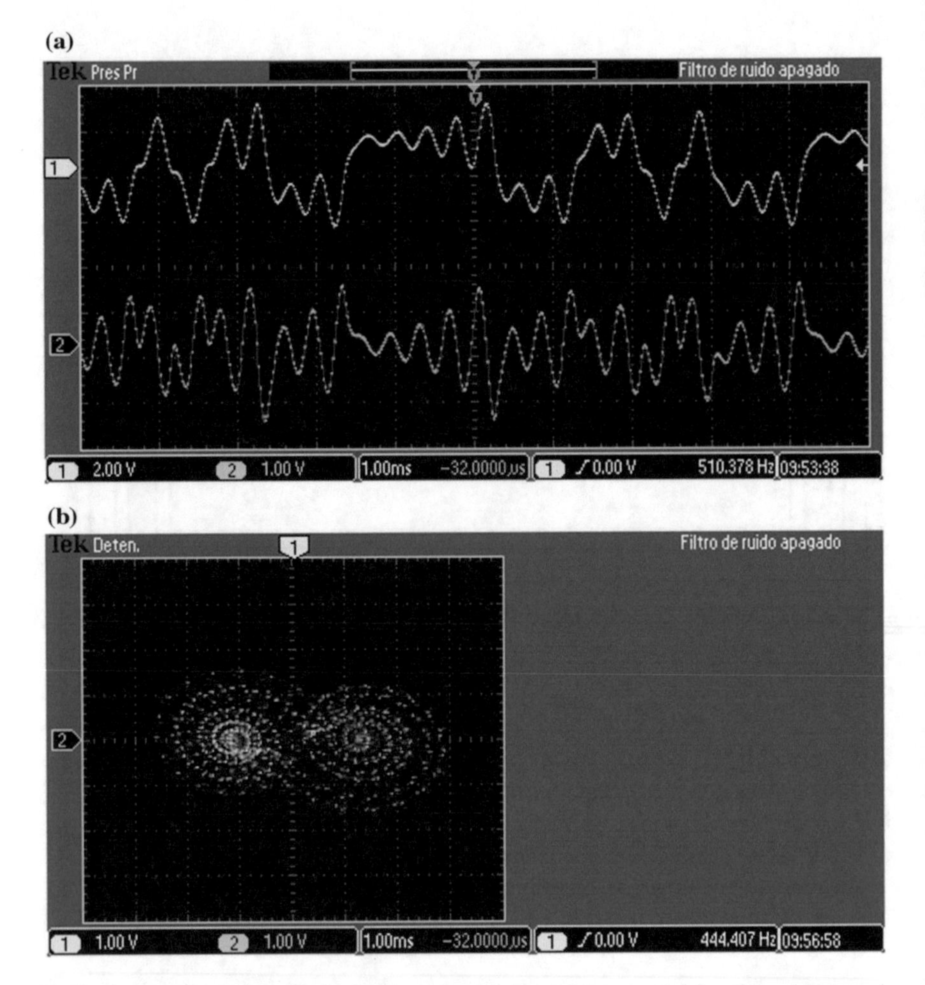


**Fig. 7.13** Experimental results for the chaos generator based on saturated functions. **a** State variables $x$ (*top*) and $y$ (*down*). **b** Portrait $x$–$y$

$$
\begin{aligned}
xm_{n+1} &= xm_n + h * ym_n \\
ym_{n+1} &= ym_n + h * zm_n \\
zm_{n+1} &= zm_n + h * (-a * x - b * y - c * z + d_1 * f(xm_n; q))
\end{aligned}
\tag{7.36}
$$

$$
\begin{aligned}
xs_{n+1} &= xs_n + h * (ym_n + k_1 * (xm_n - xs_n)) \\
ys_{n+1} &= ys_n + h * (zm_n + k_2 * (xm_n - xs_n)) \\
zs_{n+1} &= zs_n + h * (-a * x - b * y - c * z + d_1 * f(xs_n; q) + k_3 * (xm_n - xs_n))
\end{aligned}
\tag{7.37}
$$

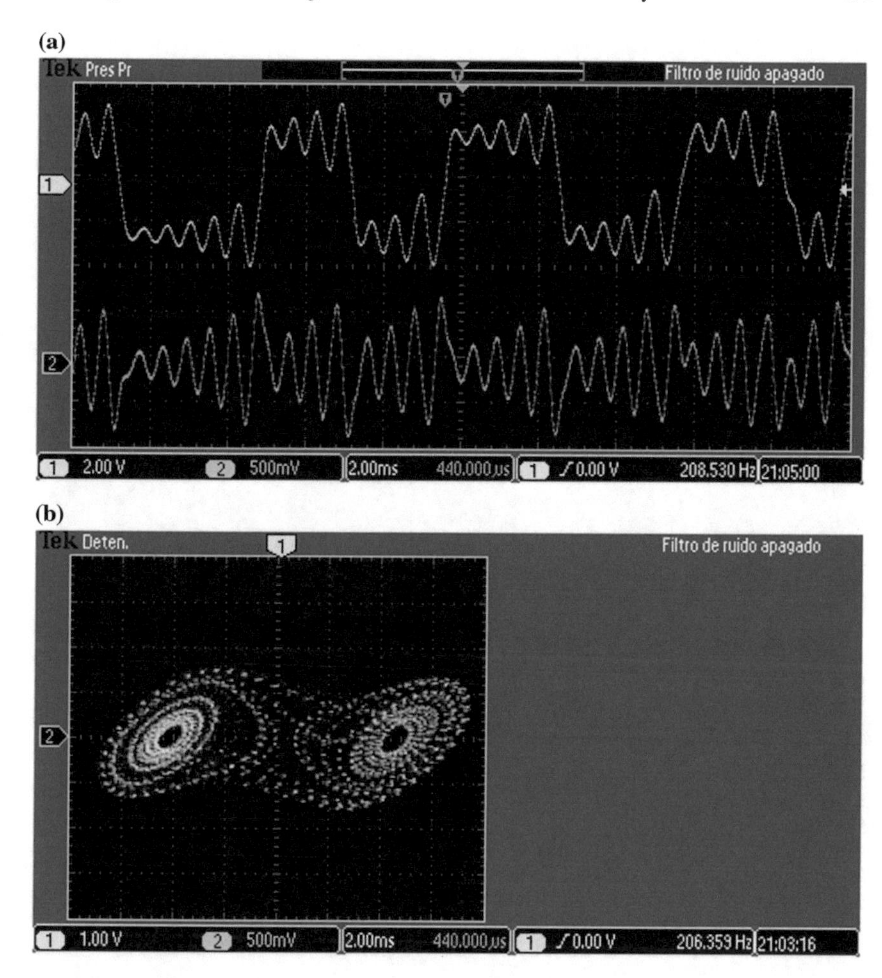


**Fig. 7.14** Experimental results for the chaos generator based on Chua's circuit. **a** State variables $x$ (*top*) and $y$ (*down*). **b** Portrait $x$–$y$

As one sees in (7.37), the multiplication $k_i * (xm_n - xs_n)$ is associated to gains $k_i$ that represent the observers for synchronizing both the master and slave chaos generators. The synchronization is controlled by evaluating the error between the state variables for the master $xm$ and slave $xs$ at each iteration $n$, i.e., $xm_n - xs_n$. Figure 7.16 shows the error evolution for two different gains of the observers, e.g., $k_i = 3$ and $k_i = 10$.

The equations in (7.36) and (7.37) should be simulated to find the appropriate gains for the observers $k_1$, $k_2$, and $k_3$. For instance, Fig. 7.17 shows the synchronization errors when the observers have the same gain and equal to 3. Figure 7.18 shows the synchronization errors but now using gains equal to 10. Following this direction, one

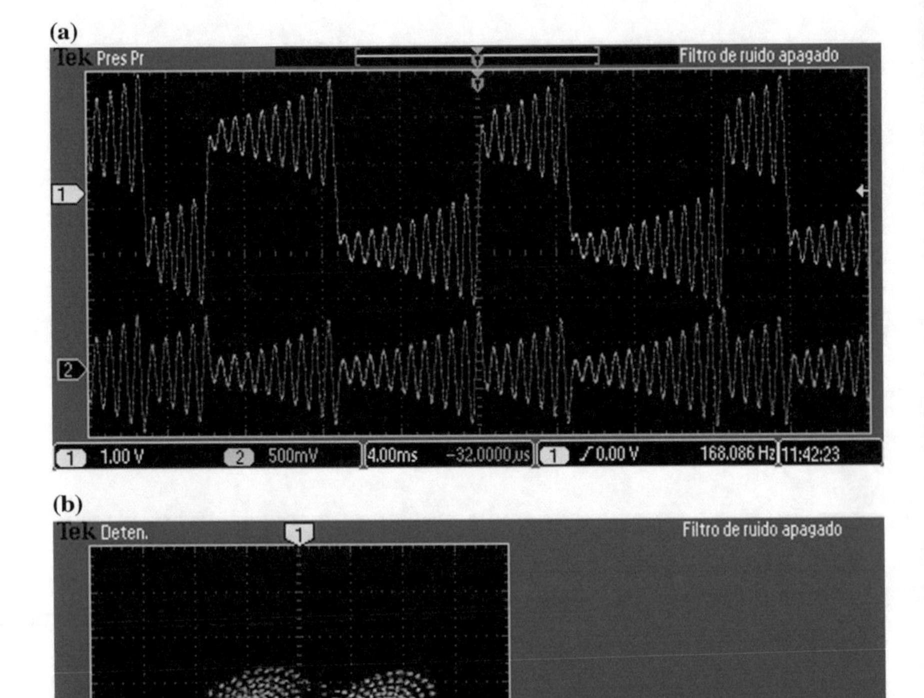


**Fig. 7.15** Experimental results for the chaos generator based on sawtooth. **a** State variables $x$ (*top*) and $y$ (*down*). **b** Portrait $x$–$y$

can conclude that the time required for synchronizing the master–slave topology is lower as the gains of the observers being increased.

The synchronization is better if two state variables are plotted and the result is a slope at 45°. For example, the plot $x_{master}$ versus $x_{slave}$ is given in Fig. 7.19, where one can appreciate that state variable $x$ perform much better than the state variables $y$ and $z$.

### *7.6.4 FPGA Realization*

The VHDL code for the master–slave synchronization is the same for multi-scroll chaotic oscillators. Figure 7.20 shows the synchronization of two multi-scroll chaotic

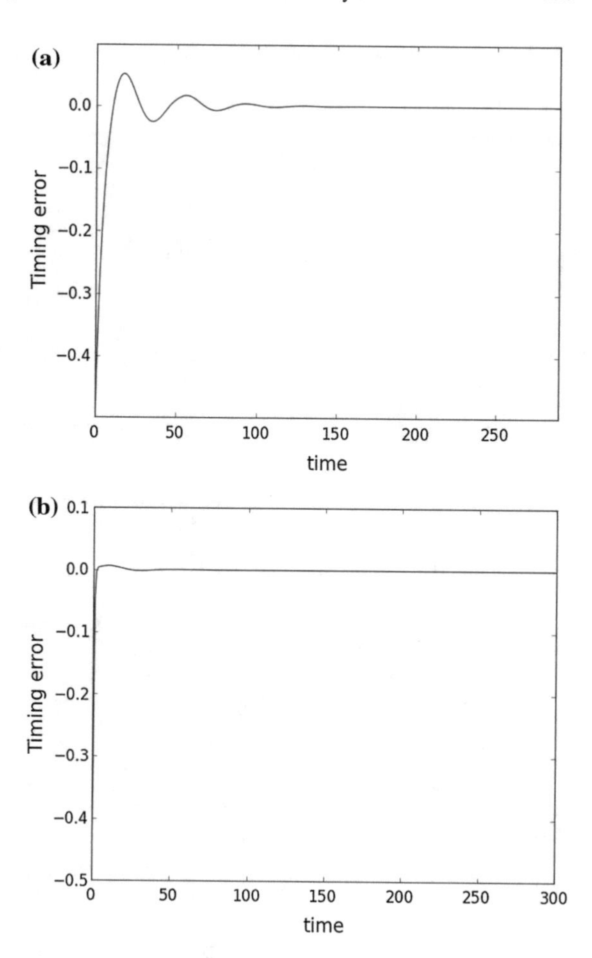


**Fig. 7.16** Synchronization error for two different gains of the observers. **a** $k_i = 3$. **b** $k_i = 10$

oscillators, where one can appreciate the master and slave blocks. As mentioned above, a multiplexer (left side) is required to process the initial conditions, afterwards, the loop is closed to iterate according to the numerical method used to solve the dynamical systems of equations. The counter controls the clock cycles to ensure the iterations and finally, the subtractors are used to provide outputs to observe the error in the three state variables in the slave. This is pretty good described in [33].

After the synchronization is guaranteed, one is able to implement a secure communication system. Recalling the classification of modulation schemes provided at the beginning of this section, this chapter highlights the one based on masking. The data being transmitted is contaminated with chaos and if perfect synchronization is achieved, at the receiver end the hidden data is extracted, i.e., chaos is subtracted from the data in the channel. To do this, several blocks are required: for example, when processing images, a read-only memory (ROM) should be realized to allocate the data, as shown in Fig. 7.21. The master oscillator requires an adder to mask the

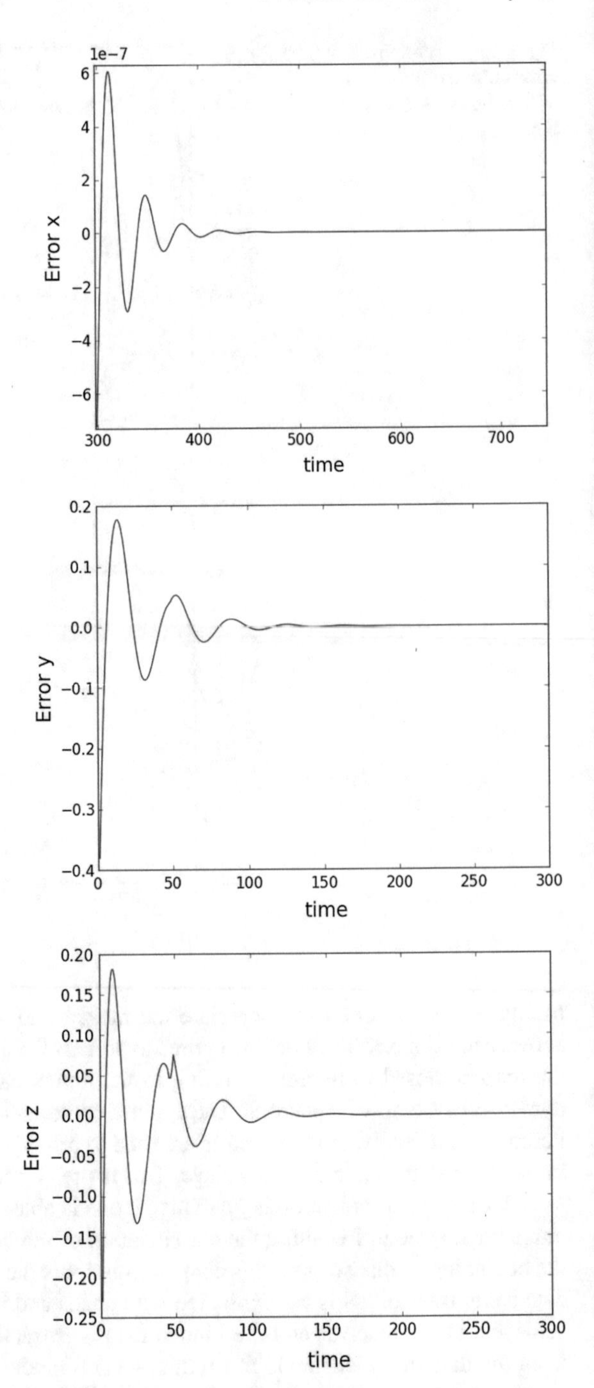


**Fig. 7.17** Synchronization errors for state variables $x$, $y$, and $z$, with $k_1 = k_2 = k_3 = 3$

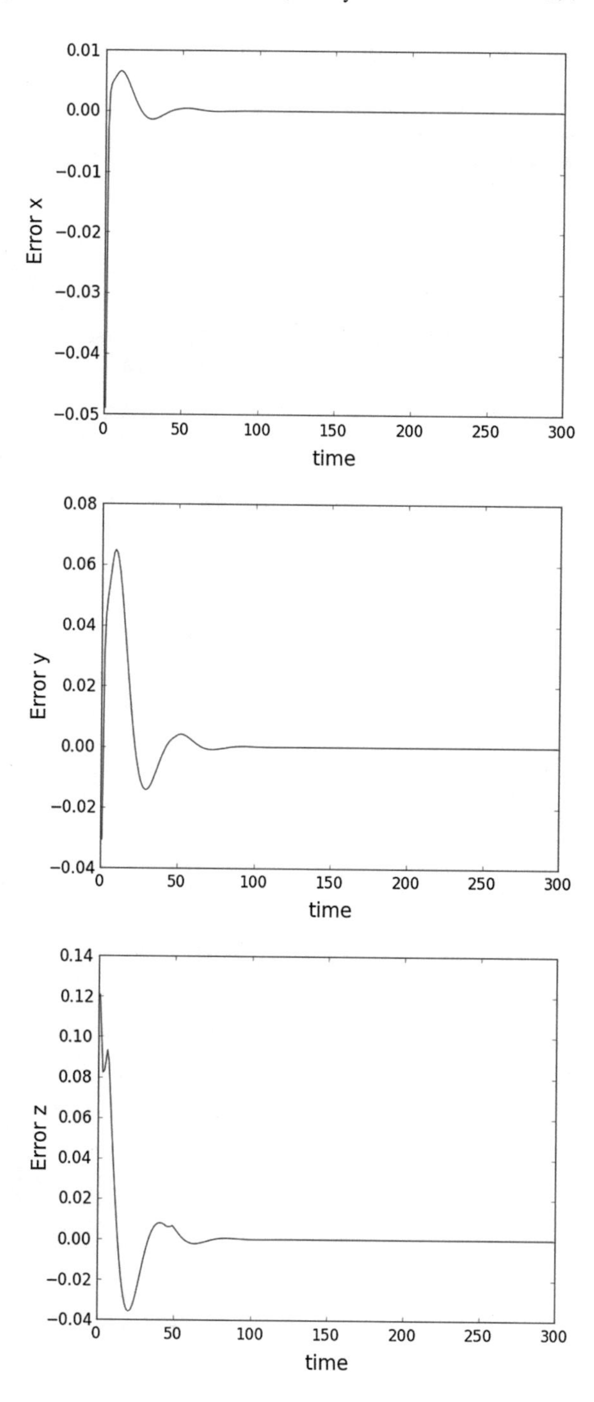


**Fig. 7.18** Synchronization errors for state variables $x$, $y$, and $z$, with $k_1 = k_2 = k_3 = 10$

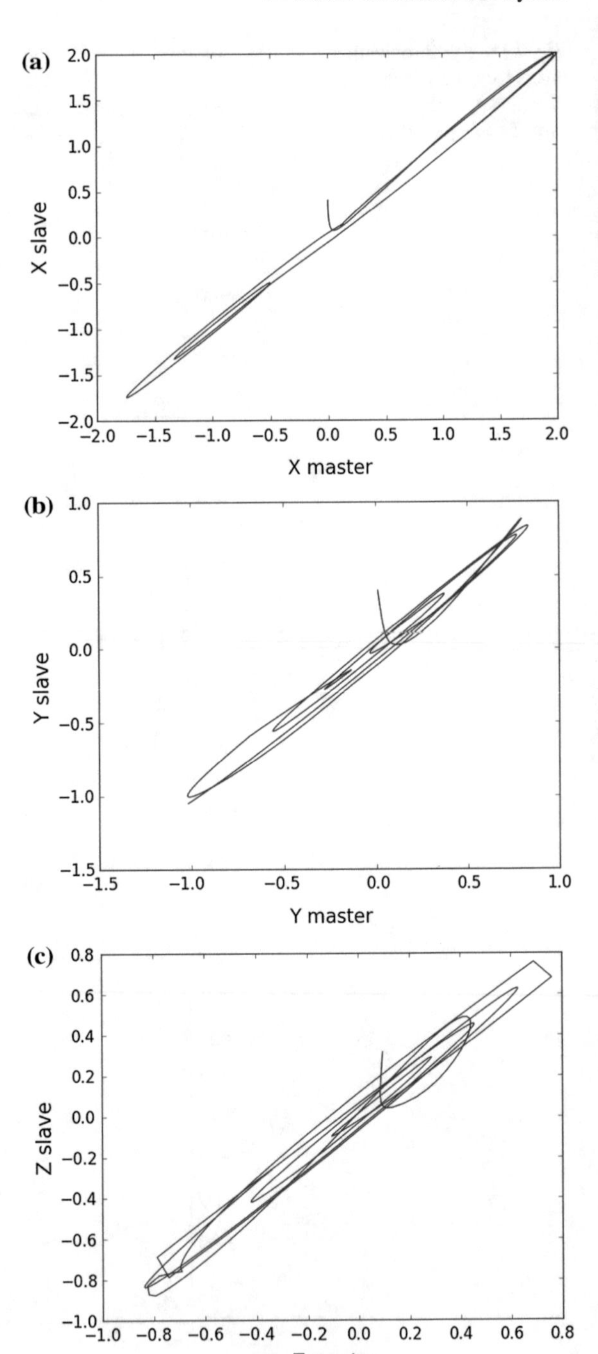


**Fig. 7.19** Error between state variables of the master and slave. **a** $x_m$ versus $x_s$. **b** $y_m$ versus $y_s$. **c** $z_m$ versus $z_s$

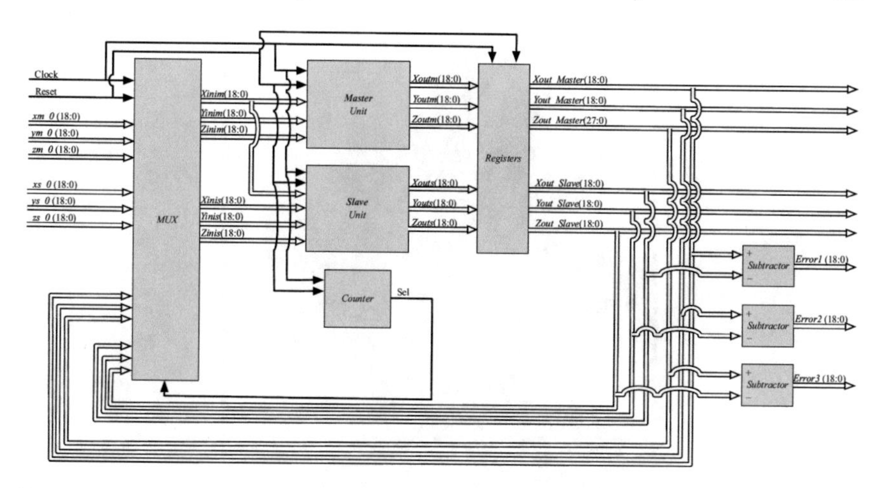


**Fig. 7.20** Block description of the master–slave synchronization and providing outputs for the errors

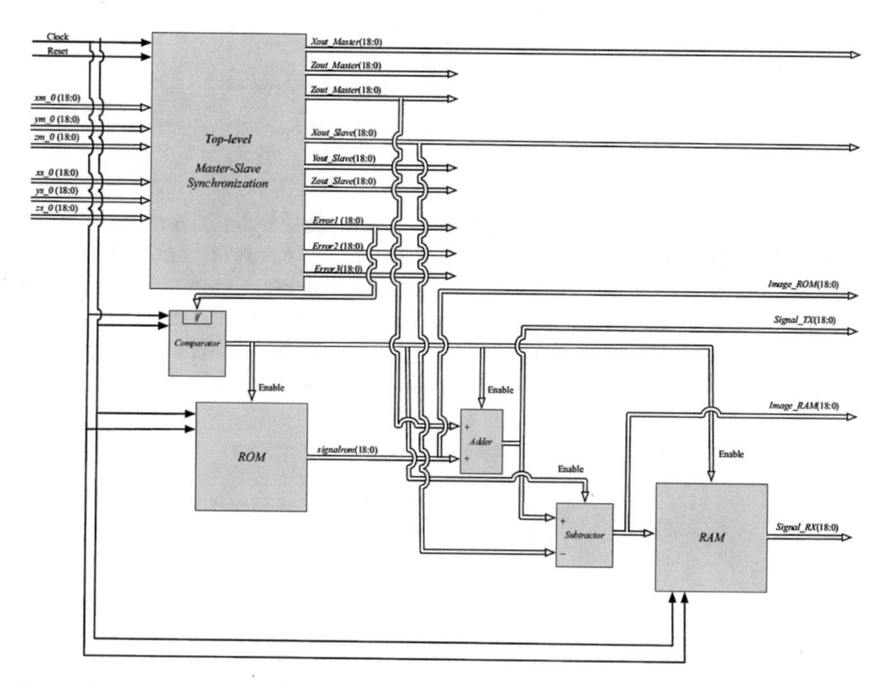


**Fig. 7.21** Chaotic secure communication system description

data to be transmitted with chaos. At the receiver end, a subtractor is required to recover the data without loss of information. The comparator is monitoring the synchronization errors, when they are lower than a required threshold, the transmitter blocks begin sending the data to the adder to encrypt it with chaos.

**Fig. 7.22** Image to be transmitted

Before the VHDL code is synthesized into an FPGA, a simulation is recommended using tools that can be freely available on Internet. For example, Active-HDL and simulink are good tools that are used herein. The data to be transmitted is the image shown in Fig. 7.22, which can be converted to a black and white or grayscale image.

The image is converted to a matrix of size (300, 226) whose values can be integers of 8 bits. Then using Simulink, the block description is given in Fig. 7.23, where one

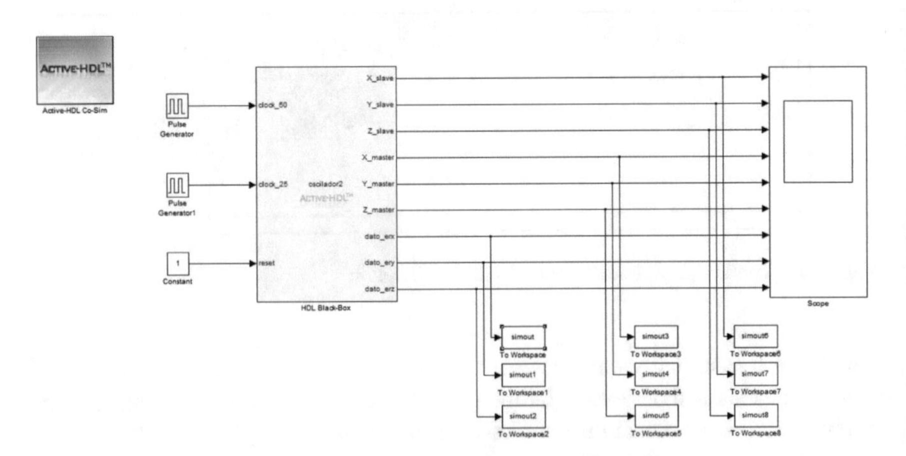


**Fig. 7.23** Simulink description for simulating the secure communication system

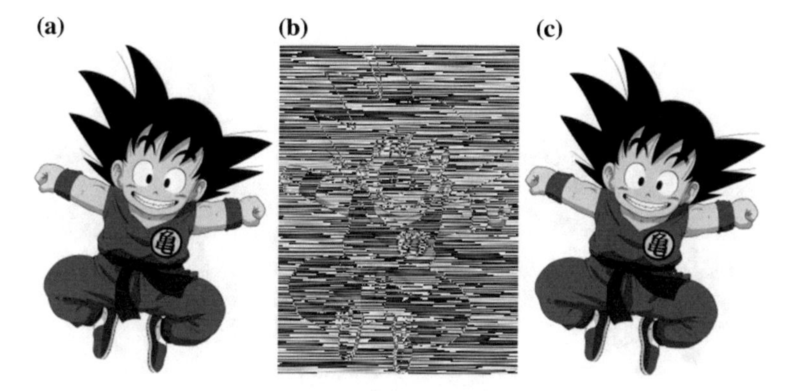


**Fig. 7.24** Image transmitted in grayscale tones using 2-scroll chaos generators. **a** Image being transmitted. **b** Channel. **c** Received image

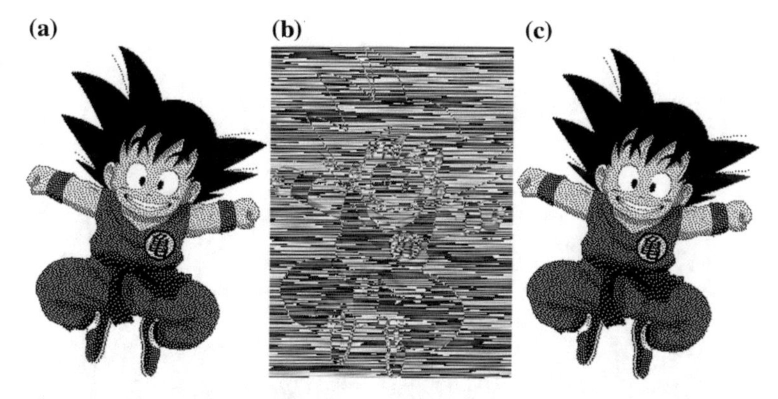


**Fig. 7.25** Image transmitted in black and white using 2-scroll chaos generators. **a** Image being transmitted. **b** Channel. **c** Received image

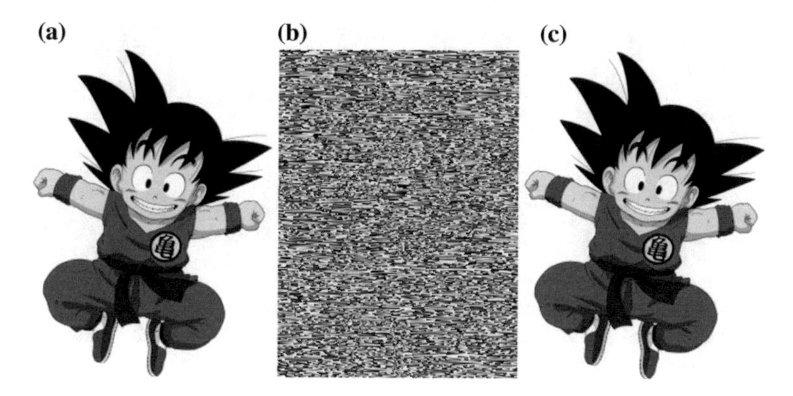


**Fig. 7.26** Image transmitted in grayscale tones using 20-scroll chaos generators. **a** Image being transmitted. **b** Channel. **c** Received image

**Fig. 7.27** Image transmitted in black and white using 20-scroll chaos generators. **a** Image being transmitted. **b** Channel. **c** Received image

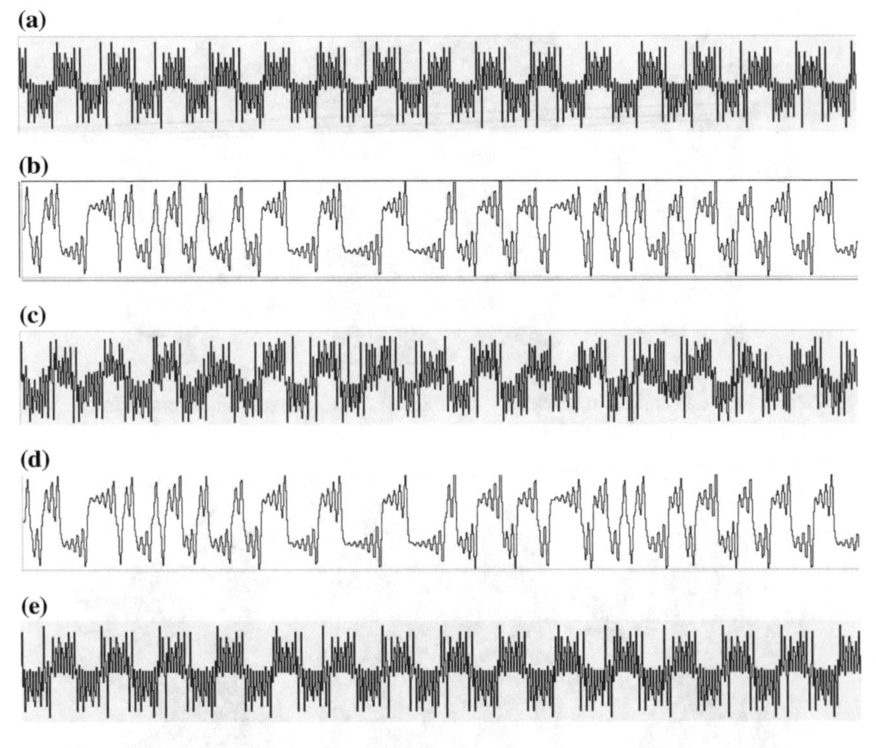


**Fig. 7.28** Signals simulated in Active-HDL for transmitting an image in grayscale tones. **a** Data being transmitted. **b** Signal in the master oscillator. **c** Channel. **d** Signal in the slave oscillator. **e** Received data

can see six outputs that are associated to the signals in the master and slave, and three outputs that provide information of the errors for the three state variables.

Using chaos generators with 2-scrolls, the experimental results for transmitting an image in grayscale tones are shown in Fig. 7.24. Figure 7.25 shows the same image transmitted in black and white.

Increasing the number of scrolls improves the unpredictability of the chaotic system. In this manner, Fig. 7.26 shows the image transmitted in grayscale tones, and Fig. 7.27 shows the image transmitted in black and white using 20-scroll chaotic oscillators.

As one sees, the channel has better unpredictability as the number of scrolls increases. However, an FPGA has limited resources and the number of scrolls being generated depend on it. The following chapter discusses some issues for the applications presented in this book, where to save hardware resources, multipliers can be implemented as single constant multipliers (SCM).

Active-HDL is also helpful to observe the information in a secure communication system. Figure 7.28 shows the signals when transmitting an image in gray scale. After one approves this simulation, the FPGA synthesis is the next step.

# Chapter 8
# Challenges in Engineering Applications

## 8.1 On the Length of the Digital Words

An implementation of a system in hardware–software, e.g., inside a FPGA, should use integer arithmetic (see Sect. 1.4) and its design should follow the following steps [92]:

1. Gather all the requirements and specifications in order to design the system efficiently.
2. Implement the algorithm in floating point arithmetic. This first algorithm can be tested on a desktop computer.
3. Estimate the range of the variables and determine the integer part.
4. Determinate the optimal fractional part by a *signal to quantization noise ratio* (SQNR) analysis.
5. Implement and test the algorithm in fixed point arithmetic, and finally,
6. Implement the hardware–software design.

Integer arithmetic is faster than the floating point arithmetic, it is also less consuming in hardware resources, it can be implemented with less number of logical gates, and thus it also requires less power consumption. Implementation of algorithms in fixed point arithmetic is not new: digital filters have been implemented in this way in at least the past 50 years [93, Chap. 9].

The power of recent computers make possible to simulate and test the design on a desktop computer before carrying the design to a FPGA.

Herein, it will be explained how a dynamical system, like the Lorenz one, can be implemented with fixed arithmetic.

### 8.1.1 Example of a Design with the Lorenz System

The Lorenz system is defined by the following three differential equations:



E. Tlelo-Cuautle et al., *Engineering Applications of FPGAs*,
DOI 10.1007/978-3-319-34115-6_8

$$\begin{aligned}\dot{x} &= \sigma(y - x),\\ \dot{y} &= x(\rho - z) - y,\\ \dot{z} &= xy - \beta z.\end{aligned} \tag{8.1}$$

Algorithm in floating point arithmetic to simulate the Lorenz system is almost straight from (8.1). Only it is necessary to integrate each variable. One can use the Euler method, which is the most simple to implement in digital hardware. It can be seen in Algorithm 1.

---

**Algorithm 1** Lorenz system in floating point arithmetic

**Require:** Values for constants $\sigma$, $\rho$, and $\beta$. Integration time step $\Delta t = 0.001$
**Require:** Initial values for $x_0$, $y_0$, and $z_0$. Simulation time $t_s$
**Ensure:** Values of $x$, $y$, and $z$

$s = \lfloor t_s/t \rfloor$
$x \leftarrow 0, y \leftarrow 0, z \leftarrow 0$
**for** $i = 1 : s$ **do**
  $f_x \leftarrow \sigma(y_0 - x_0)$
  $f_y \leftarrow x_0(\rho - z_0) - y_0$
  $f_z \leftarrow x_0 y_0 - \beta z_0$
  $x \leftarrow x + \Delta t\ f_x$ ▷ Euler integration method
  $y \leftarrow y + \Delta t\ f_y$
  $z \leftarrow z + \Delta t\ f_z$
  **print** values $x$, $y$, $z$
  $x_0 \leftarrow x$, $y_0 \leftarrow y$, $z_0 \leftarrow z$
**end for**

---

In a programming language like C, variables in Algorithm 1 must be declared as real variables (`double` in C). In a language as Python all variables are in double precision by default.

### 8.1.2 Variables Range Determination

To obtain the range for a fixed point integer implementation of Algorithm 1, first it is necessary to obtain the range of the variables by performing simulation. That way, the range is obtained by observing the graphs shown in Fig. 8.1, which was simulated for 200 s for the Lorenz system.

From the simulation graphs shown in Fig. 8.1, the following ranges for each variable are obtained:

- $x \in [-25, 25]$,
- $y \in [-30, 30]$,
- $z \in [0, 50]$.

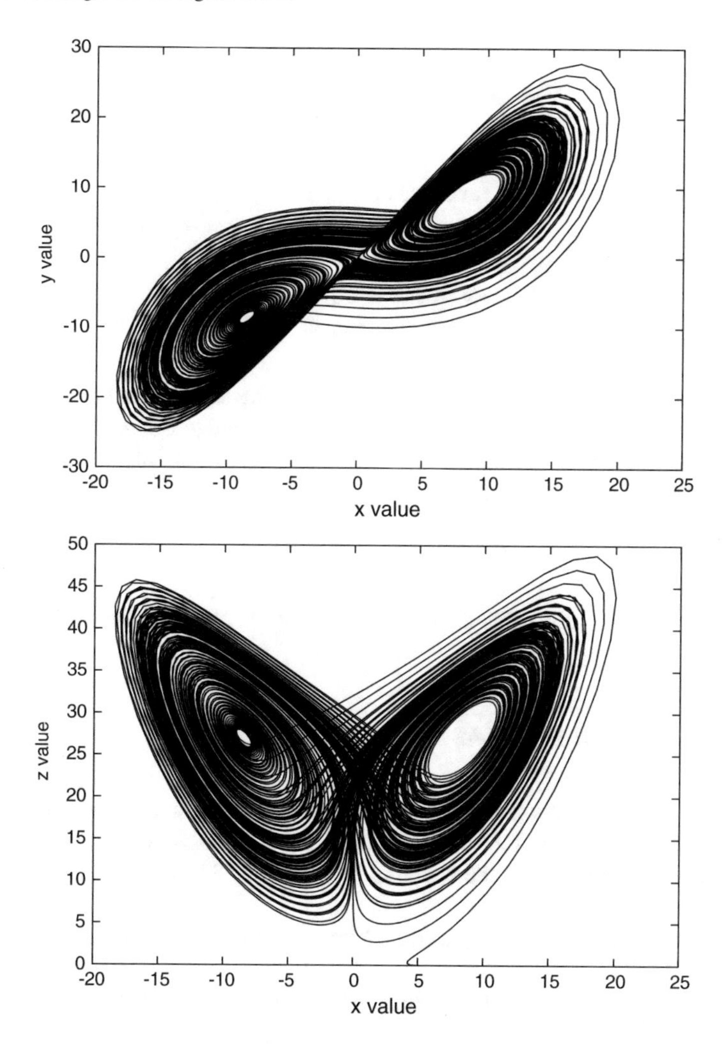


**Fig. 8.1** Simulation of 200 s for the Lorenz system. The samples are taken every 0.015 s. For this simulation $\sigma = 10$, $\rho = 28$, $\beta = 8/3$, $x_0 = 5$, $y_0 = z_0 = 0$

### 8.1.3 Number of Bits in the Integer Part

A procedure to estimate the integer part, to represent the variables of a system with integers, is by calculating the maximum values that such variables can get. It is done by replacing the maximum values in their corresponding equations. This procedure will be illustrated with the same Lorenz system and the obtained range values. Each variable is substituted with the values that produce its maximum ones as (remember that $\sigma = 10$, $\rho = 28$, $\beta = 8/3$)

$$\dot{x} = 10(30 - (-25)),$$
$$\dot{y} = 30(28 - 0) - (-30),$$
$$\dot{z} = (25)(30) - (8/3)(0).$$

the maximum values that the variables can get are:

$$\dot{x} = 10(45) = 450$$
$$\dot{y} = 840 + 30 = 870$$
$$\dot{z} = 750 = 750.$$

For the variable $\dot{y}$ it is also possible to calculate it using 50, the maximum value for $z$, then $30(28 - 50)$ which is equal to $30(-22) = -660$ and $|-660 - 30| = |-690| < |870|$, therefore, 870 is chosen because it is the biggest value that $\dot{y}$ can get.

The number of bits necessary to represent 870 is

$$\lceil \log_2(870) \rceil = \lceil 9.76 \rceil = 10.$$

Therefore, 10 + 1 bits (plus the sign bit) are need to represent for the integer part.

### 8.1.4 Fixed Point Implementation

One key question is how many bits are necessary to represent the fractional part? The first solution is to use the remaining bits of a wordlength of 32 bits, then, for 10 bits for the integer part, plus the sign bit, the remaining 21 bits could be used for the fractional part.

With numbers of type A(10, 21), it is possible to represent values in the range $-2^{10} \leq x < 2^{10} - 2^{-21}$.

The algorithm implemented in Python, using the module of deModel,[1] is presented in Algorithm 2.

Instead of deModel python module, one could use directly `long int` C variables of 64 bits; so that it is possible to represent numbers up to $A(10, 21)$, for example, which needs 32 bits. In this way also the multiplication can be stored in a 64 bit `long int` variable. Then, the 'pfx.sum' and 'pfx.substraction' can be substituted with native '+' and '−' operations.

The 'pfx.multRound' function performs the multiplication of two variables $A(a, b)$, which results in a $(2a + 1, b)$ number, and returns the rounded result to a $A(a, b)$ number. To round the result of the multiplication, it is necessary to sum a number equal to a 1 followed by $(b - 1)$ [92] zeros and then shifting the result to the right $b$ bits.

---

[1] http://www.dilloneng.com/demodel.html.

**Algorithm 2** Lorenz system in fixed point arithmetic

Values for constants $\sigma$, $\rho$, and $\beta$. Integration time $\Delta t = 0.001$
Initial values for $x_0$, $y_0$, and $z_0$. Simulation time $t_s$
Values of $x$, $y$, and $z$
$i\sigma \leftarrow A(10, 21, \sigma)$
$i\rho \leftarrow A(10, 21, \rho)$
$i\beta \leftarrow A(10, 21, -\beta)$
$s = \lfloor t_s/t \rfloor$
ix0 $\leftarrow A(10, 21, x_0)$
iy0 $\leftarrow A(10, 21, y_0)$
iz0 $\leftarrow A(10, 21, z_0)$
ix $\leftarrow A(10, 21, 0.0)$
iy $\leftarrow A(10, 21, 0.0)$
iz $\leftarrow A(10, 21, 0.0)$
ifx $\leftarrow A(10, 21, 0.0)$
ify $\leftarrow A(10, 21, 0.0)$
ifz $\leftarrow A(10, 21, 0.0)$
**for** $i = 1 : s$ **do**
  n0 $\leftarrow$ pfx.subtract( iy0, ix0 )
  ifx $\leftarrow$ pfx.multRound( i$\sigma$, n0 )
  n0 $\leftarrow$ pfx.substract( i$\rho$, iz0 )
  n1 $\leftarrow$ pfx.multRound( ix0, n0 )
  ify $\leftarrow$ pfx.substract( n1, iy0 )
  n0 $\leftarrow$ pfx.multRound( ix0, iy0 )
  n1 $\leftarrow$ pfx.multRound( i$\beta$, iz0 )
  ifz $\leftarrow$ pfx.substract( n0, n1 )
  n0 $\leftarrow$ pfx.multRound( ih, ifx ) ▷ Euler integration method
  ix $\leftarrow$ pfx.sum( ix, n0 )
  n0 $\leftarrow$ pfx.multRound( ih, ify )
  iy $\leftarrow$ pfx.sum( iy, n0 )
  n0 $\leftarrow$ pfx.multRound( ih, ifz )
  iz $\leftarrow$ pfx.sum( iz, n0 )
  **print** values float(ix), float(iy), float(iz)
  ix0 $\leftarrow$ ix, iy0 $\leftarrow$ iy, iz0 $\leftarrow$ iz
**end for**

Simulations with numbers $A(10, 21)$ and real numbers for the Lorenz system is shown in Fig. 8.2.

Which is the correct number of $b$ bits to use? In the implementation of digital filters, it is suggested to perform a *signal to quantization noise ration* (SQNR) analysis [92, 93]. But in the same form as the use of the simplest integration method (Euler method), it does not affect the behavior of the chaotic system. It could be that the values are far away of the result obtained with real numbers but still its global behavior is correct. Figure 8.3 shows this idea. The same Lorenz system in Fig. 8.2 is performed now with numbers $A(10, 19)$, $A(10, 17)$, $A(10, 15)$, $A(10, 13)$, $A(10, 11)$, and $A(10, 9)$. The last simulation, in Fig. 8.3f, stops in a single point (the one at the end of line in the center), but appears to be the other simulations are correct.

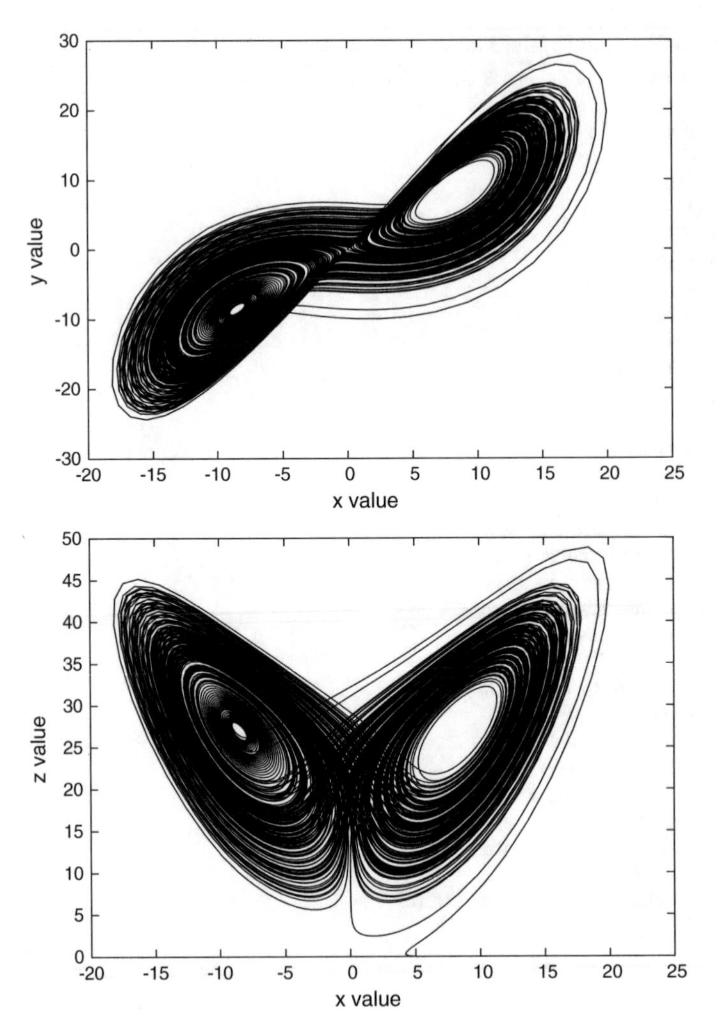


**Fig. 8.2** Simulation of 200 s for Lorenz system using fixed point arithmetic with numbers $A(10, 21)$. For the shown graphs, samples are taken every 0.015 s, and $\sigma = 10$, $\rho = 28$, $\beta = 8/3$, $x_0 = 5$, $y_0 = z_0 = 0$ for (8.1)

## 8.2 Current Challenges

As one can infer, novel research in engineering applications require combining different topics with digital hardware design. Apart from the applications provided within this book, namely: chaos generators, ANN, RNG, and secure communication systems, new challenges arise. Some related to hardware design, others related to programming issues and so on. This section lists few recent works that are of common interest to enhance the applications showed herein.

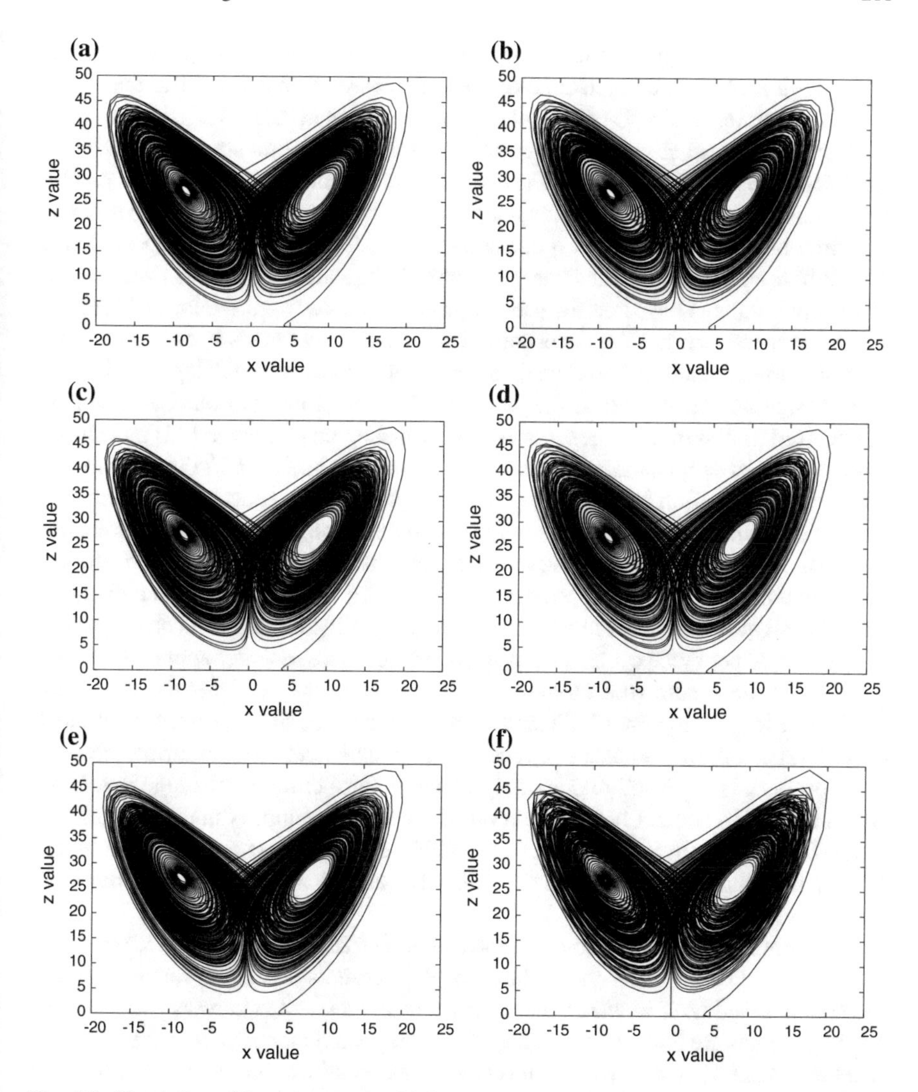


**Fig. 8.3** Simulation of Lorenz system with integer number with different size in the fractional part. The last simulation is unstable and stops in a single point (the end of the line in the center). The same parameter values in Fig. 8.2 are used. **a** Result with numbers A(10, 19). **b** Result with numbers A(10, 17). **c** Result with numbers A(10, 15). **d** Result with numbers A(10, 13). **e** Result with numbers A(10, 11). **f** Result with numbers A(10, 9)

As shown in the previous section, computer arithmetic is quite important for reliable FPGA implementations. Recently, the authors in [94] showed interest in sign-magnitude (SM) representation in decimal numbers that lies in the IEEE 754-2008 standard, and where the significant in floating-point numbers is coded as SM. The authors clearly show that software implementations do not meet performance constraints in some applications, so that more development is required in programmable logic, which is a key technology for hardware acceleration. The reader can found two strategies for SM decimal adder/subtractors in that reference, for which the authors propose six new FPGA-specific circuits. The first strategy is based on ten's complement (C10) adder/subtractors and the second one is based on parallel computation of an unsigned adder and an unsigned subtractor. Four of these alternative circuits are useful for at least one area-time-trade-off and specific operand size. The authors conclude that, the fastest SM adder/subtractor for operand sizes of 7 and 16 decimal digits is based on the second proposed strategy with delays of 3.43 and 4.33 ns, respectively, but the fastest circuit for 34-digit operands is one of the three specific implementations based on C10 adder/subtractors with a delay of 4.65 ns.

Although not mentioned in the previous chapters, a random number generator can be used to design noise generators. For instance, the authors in [95] introduce a modular design of a Gaussian noise generator (GNG) based on FPGA technology. The main contribution is the development of a new range reduction architecture in a series of elementary function evaluation modules. One issue is the approximation and quantization errors for the square root module, for which the authors showed that with a first polynomial approximation it is high; so that they avoided using the central limit theorem (CLT) to improve the noise quality. It resulted in an output rate of one sample per clock cycle. That way, the authors subsequently applied Newton's method for the square root module, thus eliminating the need for the use of the CLT because applying the CLT resulted in an output rate of two samples per clock cycle (>200 million samples per second). Two statistical tests confirmed that the GNG proposed by the authors, is of high quality. Furthermore, the range reduction, which is used to solve a limited interval of the function approximation algorithms of the System Generator platform using Xilinx FPGAs, appeared to have a higher numerical accuracy, was operated at >350 MHz, and can be suitably applied for any function evaluation.

To minimize hardware resources, look-up tables (LUT) are quite useful is some applications. For instance, the authors in [96] presents a scheme for designing a memristor-based LUT in which the memristors are connected in rows and columns. As the columns are isolated, the states of the unselected memristors in the proposed scheme are not affected by the WRITE/READ operations; therefore, the prevalent problems associated with nanocrossbars (such as the write half-select and the sneak path currents) are not encountered. In that work, the authors showed extensive simulation results with respect to the WRITE and READ operations. The performance of their proposed approach is compared with previous LUT schemes using memristors as well as SRAMs. That work showed significantly better performance in terms of WRITE time and energy dissipation for both memory operations (i.e., WRITE and READ). At the end, the authors concluded that the READ delay is nearly independent of the LUT dimension, while simulation using benchmark circuits for FPGA implementation showed that their proposed LUT offers significant improvements also at this level.

More recently, a hot topic is the application of FPGAs in biomedical applications. The work in [97] focuses on the architecture and FPGA implementation aspects of a kind of assistive tool for disabled persons with motor neuron diseases, specifically with muscle atrophy. The tool, called a communication interface, allows such persons to communicate with other people by means of moving selected muscles, e.g., within the face. To develop this project, the authors required the application of microelectromechanical system MEMS accelerometers as muscle movement sensors. In this manner, the authors investigated four different FPGA implementation methods of signal processing from MEMS sensors: manual HDL coding, usage of the Matlab HDL coder and Vivado HLS, as well as embedded microcontroller exploitation. At the end, the authors introduced a novel communication interface that can be used either as an input switch for, so called, virtual keyboards or as a stand-alone tool, which allows disabled persons to write a text by means of the Morse code.

From the current research listed above, one can imagine how to enhance engineering applications when using FPGAs. For which the software synthesis is a useful technology to accelerate the design of digital systems [98].

Readers are encouraged to search related bibliography to get more insights on the FPGA realizations for engineering applications.

# References


1. Xilinx, What is a FPGA? http://www.xilinx.com/fpga/. Accessed 05 Nov 2015
2. W. Wayne, *FPGA-Based System Design* (Pearson Education, Upper Saddle River, 2004)
3. Xilinx, *Spartan-3 FPGA Family Data Sheet*, June 2013. DS099, Product Specification
4. Altera, *Altera White Paper, FPGA Architecture*, July 2006. WP-01003-1.0
5. Altera, Logic Elements and Logic Array BLocks in the Cyclone III DEevice Family, in *Cyclone III Device Handbook*, vol. 1, December (2011)
6. Xilinx, *Spartan-3 Generation FPGA User Guide*, June 2011. UG331
7. Altera, *Cyclone IV Device Handbook*, May 2013. CYIV-51006-2.6
8. Xilinx, *Vivado Design Suite Tutorial, Hihg Level Synthesis*, April 2015. UG871
9. Altera, *Quartus II Handbook Volume 1: Design and Synthesis*, December 2014. QII5V1
10. Aldec, *Active-HDL—FPGA Simulation*, http://www.aldec.com. Accessed 02 Mar 2016
11. IEEE standard for floating-point arithmetic-redline. *IEEE Std 754-2008 (Revision of IEEE Std 754-1985) - Redline*, pp. 1–82, Aug 2008
12. Uwe Meyer-Baese, *Digital Signal Processing with Field Programmable Gate Arrays*, vol. 65 (Springer, Heidelberg, 2007)
13. V.A. Pedroni, *Circuit Design and Simulation with VHDL* (The MIT Press, Cambridge, 2010)
14. N.M. Botros, *HDL Programming Fundamentals: VHDL and Verilog (Davinci Engineering)* (Charles River Media, Inc., Newton Centre, 2005)
15. P.P. Chu, *RTL Hardware Design using VHDL: Coding for Efficiency, Portability, and Scalability* (Wiley, New York, 2006)
16. P.P. Chu, *FPGA Prototyping by VHDL Examples: Xilinx Spartan-3 Version* (Wiley, New York, 2011)
17. Aldec, *Active-HDL/Matlab-Simulink Co-Simulation Help*, http://www.aldec.com. Accessed 02 Mar 2016
18. Xilinx, *Xilinx-System Generator/Matlab-Simulink Co-Simulation Help*, http://www.xilinx.com. Accessed 02 Mar 2016
19. Altera, *DSP Builder Handbook*, http://www.altera.com, October 2015
20. Alan Wolf, Jack B. Swift, Harry L. Swinney, John A. Vastano, Determining Lyapunov exponents from a time series. Physica D: Nonlinear Phenomena **16**(3), 285–317 (1985)
21. Jinhu Lü, Guanrong Chen, Generating multiscroll chaotic attractors: theories, methods and applications. Int. J. Bifurc. Chaos **16**(04), 775–858 (2006)
22. J.M. Muoz-Pacheco, E. Tlelo-Cuautle, I. Toxqui-Toxqui, C. Snchez-Lpez, R. Trejo-Guerra, Frequency limitations in generating multi-scroll chaotic attractors using cfoas. Int. J. Electron. **101**(11), 1559–1569 (2014)
23. E. Ortega-Torres, S. Ruíz-Hernández, C. Sánchez-López, A nonlinear macromodel for current-feedback operational amplifiers. Microelectron. J. **46**(10), 941–949 (2015)




E. Tlelo-Cuautle et al., *Engineering Applications of FPGAs*,
DOI 10.1007/978-3-319-34115-6

24. C. Sánchez-López, F.V. Fernández, V.H. Carbajal-Gómez, E. Tlelo-Cuautle, J. Mendoza-López, Behavioral modeling of snfs for synthesizing multi-scroll chaotic attractors. Int. J. Nonlinear Sci. Numer. Simul. **14**(7–8), 463–469 (2013)
25. E. Tlelo-Cuautle, J.J. Rangel-Magdaleno, A.D. Pano-Azucena, P.J. Obeso-Rodelo, J.C. Nunez-Perez, FPGA realization of multi-scroll chaotic oscillators. Commun. Nonlinear Sci. Numer. Simul. **27**(13), 66–80 (2015)
26. Michael Peter Kennedy, Robust op amp realization of chua's circuit. Frequenz **46**(3–4), 66–80 (1992)
27. W.A.N.G. Fa-Qiang, L.I.U. Chong-Xin, Generation of multi-scroll chaotic attractors via the saw-tooth function. Int. J. Mod. Phys. B **22**(15), 2399–2405 (2008)
28. Z. Galias, The dangers of rounding errors for simulations and analysis of nonlinear circuits and systems?and how to avoid them. IEEE Circuits Syst. Mag. **13**(3):35–52, thirdquarter (2013)
29. J.D. Lambert, *Computational Methods in Ordinary Differential Equations* (Wiley, London, 1973)
30. R.K. Jain, Numerical solution of ordinary differential equations. Ph.D. thesis (1968)
31. Jinhu Lü, Guanrong Chen, Yu. Xinghuo, Henry Leung, Design and analysis of multiscroll chaotic attractors from saturated function series. IEEE Trans. Circuits Syst. I: Regul. Pap. **51**(12), 2476–2490 (2004)
32. L.G. de la Fraga, E. Tlelo-Cuautle, Optimizing the maximum Lyapunov exponent and phase space portraits in multi-scroll chaotic oscillators. Nonlinear Dyn. **76**(2), 1503–1515 (2014)
33. E. Tlelo-Cuautle, V.H. Carbajal-Gomez, P.J. Obeso-Rodelo, J.J. Rangel-Magdaleno, J.C. Núñez-Pérez, Fpga realization of a chaotic communication system applied to image processing. Nonlinear Dyn. **82**(4), 1879–1892 (2015)
34. V.H. Carbajal-Gomez, E. Tlelo-Cuautle, F.V. Fernandez, Optimizing the positive lyapunov exponent in multi-scroll chaotic oscillators with differential evolution algorithm. Appl. Math. Comput. **219**(15), 8163–8168 (2013)
35. D. Ruelle, *Elements of Differentiable Dynamics and Bifurcation Theory* (Elsevier, Amsterdam, 2014)
36. W. Szemplińska-Stupnicka, *Chaos Bifurcations and Fractals Around Us: A Brief Introduction*, vol. 47 (World Scientific, Singapore, 2003)
37. M.E. Yalin, Increasing the entropy of a random number generator using n-scroll chaotic attractors. Int. J. Bifurc. Chaos, **17**(12), 4471–4479 (2007)
38. K. Deb, A. Pratap, S. Agarwal, T.A.M.T. Meyarivan, A fast and elitist multiobjective genetic algorithm: NSGA-II. IEEE Trans. Evol. Comput. **6**(2), 182–197 (2002)
39. Ya.B. Pesis, Characteristic lyapunov exponents and smooth ergodic theory. Russian Math. Surveys **32**(4), 55–112 (1977)
40. R. Trejo-Guerra, E. Tlelo-Cuautle, J.M. Muñoz-Pacheco, C. Sánchez-López, C. Cruz-Hernández, On the relation between the number of scrolls and the Lyapunov exponents in PWL-functions-based n-scroll chaotic oscillators. Int. J. Nonlinear Sci. Numer. Simul. **11**, 903–910 (2010)
41. T.S. Parker, L.O. Chua, *Practical Numerical Algorithms for Chaotic Systems* (Springer, New York, 1989)
42. D. Ruelle, *Bifurcation Theory and its Application in Scientific Disciplines* (Academy of Science, New York, 1979)
43. R. Trejo-Guerra, E. Tlelo-Cuautle, J.M. Jiménez-Fuentes, C. Sánchez-López, J.M. Muñoz-Pacheco, G. Espinosa-Flores-Verdad, J.M. Rocha-Pérez, Integrated circuit generating 3-and 5-scroll attractors. Commun. Nonlinear Sci. Numer. Simul. **17**(11), 4328–4335 (2012)
44. U. Meyer-Baese. *Digital Signal Processing with Field Programmable Gate Arrays*. 3. edn (2007)
45. Malihe Molaie, Razieh Falahian, Shahriar Gharibzadeh, Sajad Jafari, Julien C. Sprott, Artificial neural networks: powerful tools for modeling chaotic behavior in the nervous system. Front. Comput. Neurosci. **8**(40), 1–3 (2014)
46. R. Chandra, Competition and collaboration in cooperative coevolution of elman recurrent neural networks for time-series prediction. IEEE Trans. Neural Netw. Learn. Syst., PP(99), 1–1 (2015)

47. Qihanyue Zhang, Xiaoping Xie, Ping Zhu, Hongping Chen, Guoguang He, Sinusoidal modulation control method in a chaotic neural network. Commun. Nonlinear Sci. Numer. Simul. **19**(8), 2793–2800 (2014)
48. Libiao Wang, Zhuo Meng, Yize Sun, Lei Guo, Mingxing Zhou, Design and analysis of a novel chaotic diagonal recurrent neural network. Commun. Nonlinear Sci. Numer. Simul. **26**(13), 11–23 (2015)
49. J. Ma, X. Song, W. Jin, C. Wang, Autapse-induced synchronization in a coupled neuronal network. Chaos, Solitons and Fractals, **80**, 31–38 (2015). (Networks of Networks)
50. Jun Ma, Xinlin Song, Jun Tang, Chunni Wang, Wave emitting and propagation induced by autapse in a forward feedback neuronal network. Neurocomputing **167**, 378–389 (2015)
51. R. Falahian, M. Mehdizadeh Dastjerdi, M. Molaie, S. Jafari, S. Gharibzadeh, Artificial neural network-based modeling of brain response to flicker light. Nonlinear Dyn. 1–17 (2015)
52. Peter Grassberger, Itamar Procaccia, Measuring the strangeness of strange attractors. Physica D: Nonlinear Phenomena **9**(12), 189–208 (1983)
53. K.G. Sheela, S.N. Deepa, Review on methods to fix number of hidden neurons in neural networks. Math. Probl. Eng. **11** (2013)
54. T. Fathima, V. Jothiprakash, Behavioural analysis of a time seriesa chaotic approach. Sadhana **39**(3), 659–676 (2014)
55. Y. Voronenko, M. Püschel, Multiplierless multiple constant multiplication. ACM Trans. Algoritm. **3**(2), (2007)
56. Y. Takahashi, T. Sekine, M. Yokoyama. A comparison of multiplierless multiple constant multiplication using common subexpression elimination method, in *51st Midwest Symposium on Circuits and Systems, 2008. MWSCAS 2008*, pp. 298–301, Aug (2008)
57. J.D.R. Antolines, Reconstruction of periodic signals using neural networks. Tecnura, **18**(01), 34–46 (2014)
58. Timothy Masters, *Practical Neural Network Recipes in C++* (Academic Press, Inc., Cambridge, 1993)
59. H.K. Kwan, Simple sigmoid-like activation function suitable for digital hardware implementation. Electron. Lett. **28**(15), 1379–1380 (1992)
60. C.-T. Lin, C.S. Lee, *Neural Fuzzy Systems: A Neuro-fuzzy Synergism to Intelligent Systems* (Prentice-Hall, Inc., Upper Saddle River, 1996)
61. L. Fausett. Fundamentals of Neural Networks: Architectures, Algorithms and Applications (1994)
62. Electron. Lett. Simple sigmoid-like activation function suitable for digital hardware implementation. **28**(15), 1379–1380 (1992)
63. L. Pang, H. Liu, B. Li, B. Liang, A data recovery method for high-speed serial communication based on FPGA. pp. 664–667, Sept (2010)
64. G.-Q. Zhong, K.-F. Man, G. Chen, A systematic approach to genererating $n$-scroll attractors. Int. J. Bifuc. Chaos **12**(12), 2907–2915 (2002)
65. L.G. de la Fraga, E. Tlelo-Cuautle, V.H. Carbajal-Gmez, J.M. Muoz-Pacheco. On Maximizing Positive Lyapunov Exponents in a Chaotic Oscillator with Heuristics. pp. 274–281, Junio (2012)
66. A. Papoulis. *Probability, Random Variables, and Stochastic Processes*, 3rd edn (McGraw-Hill, New York, 1991)
67. R. Moddemeijer, On estimation of entropy and mutual information of continuous distributions. Signal Process. **16**(3), 233–246 (1989)
68. M.E. Talçin, Increasing the entropy of a random number generator using n-scroll chaotic attractors. Int. J. Bifurc. Chaos **17**(12), 4471–4479 (2007)
69. M.B. Kennel, A.I. Mees, Context-tree modeling of observed symbolic dynamics. Phys. Rev. E, **66**, 056209–1–056209–11 (2002)
70. Y. Gao, I. Kontoyiannis, E. Bienenstock, Estimating the entropy of binary time series: methodology, some theory and a simulation study. Entropy **10**, 71–99 (2008)

71. A. Rukhin, J. Soto, J. Nechvatal, M. Smid, E. Barker, S. Leigh, M. Levenson, M. Vangel, D. Banks, A. Heckert, J. Dray, S. Vo. A statistical test suite for the validation of random number generators and pseudo random number generators for cryptographic applications, April 27, 2010. SP800-22rev1a.pdf file, last date checked: Feb 18th, 2016. Online at url: http://csrc.nist.gov/groups/ST/toolkit/rng/documentation_software.html
72. S. Ergün, A truly random number generator based on a pulse-excited cross-coupled chaotic oscillator. Comput. J. **54**(10), 1592–1602 (2011)
73. M.E. Yalçin, J.A.K. Suykens, J. Vandewalle, True random bit generation from a double-scroll attractor. IEEE Trans. Circuits Syst. **51**(7), 1395–1404 (2004)
74. R. Caponetto, L. Fortuna, S. Fazzino, M.G. Xibilia, Chaotic sequences to improve the performance of evolutionary algorithms. IEEE Tran. Evol. Comput. **7**(3), 289–304 (2003)
75. X. Wang, F.J. Hickernell, Randomized halton sequences. Math. Comput. Model. **32**(78), 887–899 (2000)
76. T. Stojanovski, J. Pihl, L. Kocarev. Chaos-based random number generatorspart ii: Practical realization. IEEE Trans. Circuits Syst. Fundam. Theory Appl. **48**(3), (2001)
77. L.M. Pecora, T.L. Carroll, Synchronization in chaotic systems. Phys. Rev. Lett. **64**(8), 821 (1990)
78. P. C. Silva, Nonlinear dynamics and chaos, From concept to application, in *Nonlinear Dynamics*, p. 2011 (2011)
79. Ashraf A. Zaher, Abdulnasser Abu-Rezq, On the design of chaos-based secure communication systems. Commun. Nonlinear Sci. Numer. Simul. **16**(9), 3721–3737 (2011)
80. Hebertt Sira-Ramirez, César Cruz-Hernández, Synchronization of chaotic systems: a generalized hamiltonian systems approach. Int. J Bifurc. Chaos **11**(05), 1381–1395 (2001)
81. S. Vaidyanathan, A.T. Azar. Anti-synchronization of identical chaotic systems using sliding mode control and an application to vaidyanathanmadhavan chaotic systems, in *Advances and Applications in Sliding Mode Control systems*, Studies in Computational Intelligence, eds. by A.T. Azar, Q. Zhu, vol. 576, pp. 527–547 (Springer International Publishing, Heidelberg, 2015)
82. S. Vaidyanathan, A.T. Azar, Analysis, control and synchronization of a nine-term 3-d novel chaotic system, in, *Chaos Modeling and Control Systems Design*, Studies in Computational Intelligence, eds. A.T. Azar, S. Vaidyanathan, vol. 581, pp. 19–38 (Springer International Publishing, Heidelberg, 2015)
83. Sundarapandian Vaidyanathan, Sivaperumal Sampath, Ahmad Taher Azar, Global chaos synchronisation of identical chaotic systems via novel sliding mode control method and its application to Zhu system. Int. J. Model., Identif. Control **23**(1), 92–100 (2015)
84. S. Vaidyanathan, B.A. Idowu, A.T Azar, Backstepping controller design for the global chaos synchronization of sprotts jerk systems, in *Chaos Modeling and Control Systems Design*, pp. 39–58 (Springer, Heidelberg, 2015)
85. S. Vaidyanathan, A.T. Azar, Analysis and control of a 4-d novel hyperchaotic system, in *Chaos Modeling and Control Systems Design*, pp. 3–17 (Springer, Heidelber, 2015)
86. S. Vaidyanathan, A.T. Azar, K. Rajagopal, P. Alexander, Design and spice implementation of a 12-term novel hyperchaotic system and its synchronisation via active control. Int. J. Model., Identif. Control **23**(3), 267–277 (2015)
87. R. Trejo-Guerra, E. Tlelo-Cuautle, C. Cruz-Hernández, C. Sánchez-López, Chaotic communication system using Chua's oscillators realized with ccii+ s. Int. J. Bifurc. Chaos **19**(12), 4217–4226 (2009)
88. M.J. Muñoz-Pacheco, E. Zambrano-Serrano, O. Félix-Beltrán, L.C. Gómez-Pavón, A. Luis-Ramos, Synchronization of pwl function-based 2d and 3d multi-scroll chaotic systems. Nonlinear Dyn. **70**(2), 1633–1643 (2012)
89. V.H. Carbajal-Gomez, E. Tlelo-Cuautle, R. Trejo-Guerra, J.M. Muñoz-Pacheco, Simulating the synchronization of multi-scroll chaotic oscillators, in *IEEE International Symposium on Circuits and Systems (ISCAS), 2013*, pp. 1773–1776 (IEEE, New York, 2013)
90. R. Trejo-Guerra, E. Tlelo-Cuautle, C. Sánchez-López, J.M. Munoz-Pacheco, C. Cruz-Hernández, Realization of multiscroll chaotic attractors by using current-feedback operational amplifiers. Revista mexicana de física **56**(4), 268–274 (2010)

91. J.M. Muñoz-Pacheco, E. Zambrano-Serrano, O.G. Félix-Beltrán, E. Tlelo-Cuautle, L.C. Gómez-Pavón, R. Trejo-Guerra, A. Luis-Ramos, C. Sánchez-López, Selected topics in nonlinear dynamics and theoretical electrical engineering, in *Chapter On the Synchronization of 1D and 2D Multi-scroll Chaotic Oscillators*, pp. 19–40 (Springer, Heidelberg, 2013)
92. S.A. Khan, Digital design of signal processing systems: a practical approach, in *Chapter System Design Flow and Fixed-point Arithmetic* (Wiley, New York, 2011)
93. A.V. Oppenheim, R.W. Schafer, *Digital Signal Processing* (Prentice Hall, Upper Saddle River, 1975)
94. M. Vázquez, E. Todorovich, FPGA-specific decimal sign-magnitude addition and subtraction. Int. J. Electron. **103**(7), 1166–1185 (2016)
95. Yuan-Ping Li, Ta-Sung Lee, Jeng-Kuang Hwang, Modular design and implementation of field-programmable-gate-array-based gaussian noise generator. Int. J. Electron. **103**(5), 819–830 (2016)
96. T.N. Kumar, H.A.F. Almurib, F. Lombardi, Design of a memristor-based look-up table (lut) for low-energy operation of FPGAs. Integr. VLSI J. **55**, 1–11 (2016)
97. Zbigniew Hajduk, FPGA-based communication interface for persons with motor neuron diseases. Biomed. Signal Process. Control **27**, 51–59 (2016)
98. Jian Wang, Yubai Li, A novel model of computation for software synthesis based on data frame driving. IETE Tech. Rev. **32**(1), 70–78 (2015)

# Index




E. Tlelo-Cuautle et al., *Engineering Applications of FPGAs*,
DOI 10.1007/978-3-319-34115-6

Printed in the United States
By Bookmasters